HIGHER
Product Design

grade **booster**

Mark Leishman ✗ David McMillan

For Kara, Craig and Sam

03/201011

ISBN 978-1-84372-477-3

Published by
Leckie & Leckie Ltd
An imprint of HarperCollins*Publishers*
Westerhill Road, Bishopbriggs, Glasgow G64 2QT
T: 0844 576 8126 F: 0844 576 8131
leckieandleckie@harpercollins.co.uk www.leckieandleckie.co.uk

Special thanks to
Susan McLaren (content review), Roda Morrison (copyedit),
The Partnership Publishing Solutions Ltd (design and page make-up),
Tara Watson (proofreading), Caleb Rutherford (cover design).

A CIP Catalogue record for this book is available from the British Library.

Leckie & Leckie is grateful to the following for their permission to reproduce their material:
The Scottish Qualifications Authority for permission to reproduce past examination questions;
Christies's images/CORBIS (p20); Jon Arnold/Alamy (p20); Molly Curtin (p24); Comstock Images/Alamy (p27 & 30); Philip Scalia/Alamy (p28); Steve Cadman (p29).

FSC™ is a non-profit international organisation established to promote the responsible management of the world's forests. Products carrying the FSC label are independently certified to assure consumers that they come from forests that are managed to meet the social, economic and ecological needs of present and future generations, and other controlled sources.

Find out more about HarperCollins and the environment at
www.harpercollins.co.uk/green

CONTENTS

Skill Builders

1 Introduction

- ***When should I use this book?***
- ***How should I use this book?***
- ***Preparing for assessment***
- ***Higher Product Design: rationale***
- ***Course structure***
- ***Assessment***
- ***Having a study plan***
- ***Thinking about applying to university?***
- ***Entrance qualifications***
- ***Portfolio***
- ***Calculating your UCAS points***
- ***University courses***

This book has been written to support your studies in Higher Product Design. There are lots of activities and exam questions for you to try as well as valuable tips on how to achieve success and gain top grades. There is also some good advice on how to prepare for the final exam and suggestions on how you could continue your studies once you leave school.

WHEN SHOULD I USE THIS BOOK?

You can use this book at any time throughout your course. It is a resource that you can dip in and out of.

The chapter on design history gives a very brief overview of product design over the last century and should help you to put this course into context. It may also inspire your own design work and help you to develop your own style. Each of the seven styles and periods identified provides you with examples of products and the names of people associated with that style. This will give you a starting point for research if you want to find out more.

The chapter on the design assignment should be read before you begin your own design assignment which will be sent to the Scottish Qualifications Authority (SQA) in April. The assignment should be given to you around February or March. Chapter 4 will help you to prepare for the written exam which you will sit some time in June. Work on this can begin as soon as you start covering the topics listed on pages 78–82 in class.

HOW SHOULD I USE THIS BOOK?

This will really depend on you. For example you may want to try to write the answers to the questions in chapter 4 yourself before comparing them with the sample answers given or you may prefer just to read the pupil answers and consider the advice given on how these pupils got their marks. You can begin to prepare for the design assignment any time. There is a sample design assignment for you to try as well as skill-builder exercises.

This book is not intended to be read from cover to cover in one go. You should familiarise yourself with the content and where things are, and then use it as you need to.

PREPARING FOR ASSESSMENT

This book has been written to help you prepare for assessment in Higher Product Design but it does not give you the depth of knowledge and skill necessary to pass these assessments. The course content is covered in another book called *Higher Product Design Course Notes* and is also published by Leckie and Leckie. Of course you also need to listen very carefully to your teacher and to what is said in class, as well as practising the skills needed to carry out your own design work.

HIGHER PRODUCT DESIGN: RATIONALE

The following extracts have been taken from the rationale in the course specification for Higher Product Design produced by the SQA.

> *'This course will help develop creative flexible learners who are able to work autonomously, to achieve good quality, feasible proposals or outcomes through active experiences of product design. At its heart is creativity. The course develops the ability to apply skills and knowledge in different situations – attributes which are becoming more and more valuable to individuals and organisations'.*
>
> *'Candidates undertaking this course will be in a strong position to pursue further study in all areas of design and manufacturing. The course will also contribute to personal development, augmenting transferable skills which will be useful regardless of the career path followed'.*
>
> *'Manufacturing industry remains the cornerstone of the Scottish economy. Any decline poses a threat to the health, sustainability and diversity of the country and therefore to its infrastructure and the prosperity of its citizens ... because of its diverse, dynamic nature, manufacturing needs people with equally diverse and adaptable skills.'*
>
> *Make it in Scotland* (www.makeitinscotland.co.uk)

COURSE STRUCTURE

The course is divided into three units. This does not mean that your school or centre will teach the content of these units sequentially. Each centre will have its own course structure and methods of teaching. It is important that you listen to your teachers and take their advice. They will have devised a course with assignments and homework to enable you to meet all the assessment requirements necessary. The content of each of the units is outlined below.

Unit (a): Design Analysis

1. Evaluate a commercial product.
   - Aspects to be included in the evaluation are identified and justified.
   - An appropriate strategy for evaluation is developed.
   - A comprehensive evaluation of the product is carried out.
   - Valid conclusions about the product are given.

2. Establish a specification from a brief.
   - The brief is analysed and relevant design issues are identified.
   - The design issues are fully researched.
   - A detailed specification is derived from the design issues researched.

Unit (b): Developing Design Proposals

1. Produce a design proposal
   - A wide range of alternative ideas is generated and developed.
   - A design proposal is reached through the application of design knowledge.
   - Decisions made in reaching the design proposal are recorded and justified.
2. Use graphic techniques during the production of a design proposal.
   - A range of types of drawings and sketches is produced.
   - Drawings and sketches are used to effectively communicate the development of ideas and the design proposal.
   - Rendering skills are used to effectively communicate the development of ideas and the design proposal.
3. Use modelling techniques during the production of a design proposal.
   - A range of types of model is produced.
   - Practical skills are used effectively.

Unit (c): Manufacturing Products

1. Explain why particular materials are used for the manufacture of commercial products.
   - Materials used to manufacture given products are correctly identified.
   - Valid reasons are given to justify the identification of materials.
   - A valid explanation is given of why materials are suitable for the given products.
2. Explain why particular processes and systems are used for the manufacture of commercial products.
   - Processes and systems used to manufacture given products are correctly identified.
   - Valid reasons are given to justify the identification of processes.
   - A valid explanation is given of why processes are suitable for the given products.

3. Produce an orthographic drawing suitable for use in the manufacture of a given product.
   - A complete and accurate orthographic drawing is produced.
   - Line type and dimensioning are correct.

ASSESSMENT

The three units outlined above will be assessed by a combination of written tests and course work. Your teacher will mark all of your course work using a standard set by a National Assessment Bank (NAB). This ensures that standards in all schools and presenting centres are the same. If you pass all the assessments for the units outlined above then you will be presented for the external examination which is divided into two parts.

Part 1	Design assignment	70 marks
Part 2	2-hour written exam	70 marks

Your final grade for Higher Product Design will be based on the combined mark of these two parts.

HAVING A STUDY PLAN

It is important that you begin to organise and plan your learning as early as possible so that you are prepared and ready for all of your assessments. It is likely that Higher Product Design will be only one of a group of subjects that you are studying and that you will have other subjects to revise for too. This makes it even more essential that you have a plan and a structure that allows you to complete all of your revision in time, allowing you to feel confident and prepared.

Exams are a stressful time for everyone and you can minimise that stress in the lead up to the exam by writing down what you have to revise and then committing yourself to setting aside time to do that revision.

You are an individual

You must first of all recognise that you are different from everyone else. The demands placed on your time will be different to those placed on your friends. You may have a commitment to go to a sports club twice a week, you may have a part time job or have a home routine that involves a commitment to other members of your family. It may be that you meet with friends regularly, or enjoy watching a television programme at a set time each evening. All of these things are a necessary part of your life and it is important that you do not give it all up in order to study. You need to be able to relax, take regular exercise and have fun. These social activities that we all enjoy will actually help you to concentrate and learn better. However, revising for exams is important too and you may have to make some sacrifices to ensure that you give yourself enough time to revise properly and understand what you have learnt. You must make a commitment to yourself to study and revise thoroughly.

Finding time

One of the biggest mistakes made by pupils studying for exams at school is that they leave everything to the last minute only to discover that they do not have enough time to revise properly. You need to sit down now and identify how much available time you have. To do this you should draw up a table using hourly time slots and map out what a typical week is like for you at the moment. You can then identify specific time slots in the week where you are free to study.

What needs to be revised?

Once you have identified when you are able to study, you need to know what to study. If you identify five single hour time slots to study Product Design you need to know specifically what topics to study each hour and know that you have enough time before your exam to study all the topics required in sufficient depth. The table opposite uses the nine topics for revision identified on pages 78– 82. Week one is shown and you will see that two topics were revised this week. If a similar study pattern is followed each week then it will take five weeks to revise all nine topics. It may then be useful to recap on all that revision, taking on a topic for each hour over a further two weeks.

This study plan means that you should begin a structured revision programme at least seven weeks before your exam. If you begin your study three weeks before your exam then you will need to spend ten to twelve hours a week on Product Design, which does not leave a lot of time for other subjects. This is called cramming and it is not an effective way to revise.

Each week a different two topics would be revised for Product Design. Revision strategies may include reading your text book, making new notes, reading existing notes and doing past papers.

This is a busy schedule for a pupil who is sitting four highers. Obviously once exam leave starts there will be more time to consolidate learning and revision, but that is too late to begin. This timetable for revision should begin about eight weeks before your first exam. Once written you must commit to it. There is lots of free time built in and it is important to have a plan for your free time as well. This will help you relax and take your mind off things. Don't sit around doing nothing. Be active, meet your friends or do something together as a family. You will also find that if you stick to your plan you will feel that you are making progress and are in control. This is a good feeling and you will be able to enjoy your free time much more.

	Mon	Tues	Wed	Thurs	Fri	Sat	Sun
08.00							
09.00	School	School	School	School	School		
10.00	School	School	School	School	School	Maths	Design for people
11.00	School	School	School	School	School	Biology	
12.00	School	School	School	School	School	Biology	English
13.00	School	School	School	School			
14.00	School	School	School	School			Maths
15.00	School	School	School	School	Design for people		
16.00	Biology						Biology
17.00		English	Biology		English		
18.00	Manuf Systems		Youth Club	Maths	Maths		
19.00		Sports Club	Youth Club				Product Design Summary
20.00	English	Sports Club	Youth Club	Maths			
21.00		Manuf Systems	English				
22.00							

You need to devise a study plan that works for you. During your course work you may not study quite as much, but you should spend at least two hours every week on each subject. Revision is an all year round activity that will intensify as your prelims and final exams get nearer. If you get into good habits now, it will benefit you in the end.

THINKING ABOUT APPLYING TO UNIVERSITY?

When considering which university to apply for, the two most important decisions you will have to make are what you want to study and where you want to go. Design and Technology courses are on offer throughout Britain and a selection of the types of courses on offer are listed on pages 14 to 16.

Before making your decision you should:

- ask advice from teachers, careers advisers, family and friends
- attend higher education conferences and university open days
- read the university prospectus and course descriptor.

Useful websites

- www.ucas.ac.uk
- www.springboard.co.uk
- www.hero.ac.uk

ENTRANCE QUALIFICATIONS

Once you have decided which course you wish to study you need to find out what the entrance qualifications are. Most courses will specify a subject or subjects that you must have studied successfully at school. For example, you may need to have passed Higher Grade English or Higher Grade Maths. The university will then ask for minimum entry requirements and an offer of a place may be based on the grades you have in other subjects or a points tally.

A university may ask applicants for a certain number of UCAS points before being offered a place on a course. This points tally establishes agreed equivalences between different types of qualifications and provides comparisons between different applicants with different types of qualifications. A points offer means that each qualification you have will be awarded a points score. Together they will give you a points tally which must at least meet the points allocated to your chosen course. All Standard Grades, Highers and Advanced Highers have a points score allocated to them. These are listed on the page opposite.

A grades offer requires a minimum standard of grades, usually in a specified subject. For example a B pass in English may be specified along with a B and two Cs in any of the other Highers you sit. If you do not meet these minimum grades, you will be rejected regardless of your points total.

PORTFOLIO

Many design-based courses also ask for a portfolio of drawings and design work. This is an important part of the admissions process. You should compile examples of your best work which can be drawn from Art and Design, Product Design and Graphic Communication. You should also include photographs of three dimensional work you have made including models, prototypes and finished products.

CALCULATING YOUR UCAS POINTS

Adv. Higher	Higher	Int 2	Standard Grade	Points
A				120
B				100
				90
C				80
				77
D	A			72
				71
				64
	B			60
				58
				52
				50
	C			48
				45
	D	A		42
				39
			Grade 1	38
		B		35
				33
		C	Grade 2	28
				26
				20
				14
				7

You may also be given credit for other qualifications that you hold such as music examinations and speech and drama examinations. It is worth checking if any other qualifications you have will contribute to your points tally.

There is no double counting – applicants cannot count the same or similar qualifications twice. Achievement at a lower level will be subsumed into the higher level. This means that you will not be given credit for a Standard Grade if you have passed the Higher.

UNIVERSITY COURSES

There are many university courses on offer which will allow you to continue to study design once you complete this course. A selection are shown below which cover Scotland and different parts of England. They have been chosen to give an indication of the range and type of course on offer. Many more are available throughout the United Kingdom.

Telford College Edinburgh **www.ed-coll.ac.uk**

- HND Interior Design
- HND Graphic and Digital Design

Northumbria University **www.northumbria.ac.uk**

- Product Design and Technology BSc (Hons)
- Mechanical Design and Technology BEng (Hons)
- Design for Industry BA (Hons)
- Interior Design BA (Hons)
- Three Dimensional Design: Furniture and Product BA (Hons)
- Transportation Design BA (Hons)
- Architectural Design and Management BA (Hons)
- Architectural Technology BSc (Hons)

Robert Gordon University Aberdeen **www.rgu.ac.uk**

- MArch/BSc (Hons) Architecture
- MArch/BSc (Hons) Interior Architecture
- BSc (Hons) Architectural Technology
- BA (Hons) Three Dimensional Design (Jewellery and Ceramics)
- BDes (Hons) Product Design
- MEng Mechanical Engineering

University of Dundee www.dundee.ac.uk

- MArch/BA (Hons) Architecture
- BDes Design (Hons)
- BDes Interior and Environmental Design
- BSc (Hons) Innovative Product Design

Brunel University (West London) www.brunel.ac.uk

- Industrial Design BSc
- Industrial Design and Technology BA
- Product Design BSc
- Product Design and Engineering BSc
- Virtual Product Design BSc
- Design for Sport BSc

Edinburgh College of Art www.eca.ac.uk

- BA (Hons) Design and Applied Arts
- BArch (Hons) Architecture
- MA (Hons) Landscape Architecture
- BSc (Hons) Integrated Product Design

Glasgow School of Art www.gsa.ac.uk

- BA (Hons) Design
- BDes (Hons)/MEDes Product Design
- BEng/MEng Product Design Engineering
- BArch (Hons) Architectural Studies
- Dip Arch (Diploma in Architecture)

Napier University Edinburgh www.napier.ac.uk

- Consumer Product Design BDes (Hons)
- Design Futures BDes (Hons)
- Interdisciplinary Design MDes
- Interior Architecture BDes (Hons)
- Civil Engineering BSc (Hons)

Coventry University www.coventry.ac.uk

- Vehicle Design MDes
- Sports Product Design MDes
- Intelligent Product Design MDes/BA (Hons)
- Industrial Product Design MDes/BSc (Hons)
- Consumer Product Design MDes
- Computer Aided Product Design MDes/BSc (Hons)
- Transport Design Futures MDes
- Architectural Design and Technology BSc (Hons)
- Applied Arts BA (Hons)

Adam Smith College Fife **www.fife.ac.uk**

- HND Furniture Construction and Design
- PDH Advanced Diploma Interior Design
- HND Computer Aided Draughting and Design
- HND Architectural Technology
- HND Quantity Surveying
- HND Construction Management

University of Huddersfield **www.hud.ac.uk**

- Product Design BA/ BSc (Hons)
- Product Design: Children's Products and Toys BA(Hons)
- Product Design: 3D Animation BA (Hons)
- Product Design: Sustainable Design BA (Hons)
- Product Innovation Design and Development BSc (Hons)
- Transport Design BA (Hons)
- Interior Design BA (Hons)
- Exhibition and Retail Design BA (Hons)
- Automotive Design and Technology BSc (Hons)

Strathclyde University **www.strath.ac.uk**

- BEng / MEng Product Design Engineering
- BSc (Hons) Product Innovation
- BEng Sports Engineering
- BArch (Hons) Architecture

2 Design History

- ***Introduction to design history***
- ***Art Nouveau***
- ***Charles Rennie Mackintosh***
- ***De Stijl***
- ***Bauhaus***
- ***Art Deco***
- ***1950s and 1960s Style***
- ***Postmodernism***

INTRODUCTION TO DESIGN HISTORY

The history of design can be traced back to the writings of the architect and military engineer Vitruvius (80–10BC). In his writings he established a principle that applies equally to buildings and products today: 'all buildings must satisfy three criteria: strength, functionality and beauty'. These principles were embraced by the Modernists who wrote that 'form follows function' and by all designers to a greater or lesser extent.

Learning about the history of product design and examining the work of well-known designers will help you to develop your own style. You may find that viewing other people's work will inspire you and help you to understand many of the complex issues surrounding product design today.

This short chapter on design history is only an introduction to a topic that has been written about extensively over the years. The design movements, styles and names contained in each of the seven short commentaries can act as a starting point for further research. There is a lot of information that you can access through the Internet and libraries and you can begin to build up your own picture and information reference for the styles and designs that interest you.

There are many other design styles that you may also find interesting, such as the Arts and Crafts Movement, the Shakers and Constructivism. There are designers not mentioned in this book who, in their own ways, have influenced product design and many of the products that we use today, such as Terence Conran, Victor Papanek, Henry Ford and James Dyson. There are also companies, such as Braun, Apple, Sony and Olivetti, which have designed and manufactured products that have revolutionised lifestyles.

It is important for you to begin to examine everyday products as well as those that are regarded as design classics. Find out how they are made and what materials they are made from. Discuss with your teacher and your friends who you think a particular product is targeted at. Talk about the aesthetics of the product and consider how the product might be improved.

The products that we use today are now part of design history. It is the job of the next generation of designers to design and develop the products that will shape our future.

ART NOUVEAU

Art Nouveau is the name given to a style of design popular in Europe and America between 1890 and 1914. It was first seen in the work displayed in L'Art Nouveau, a gallery in Paris from which the name was taken.

It is a frequent misconception and over simplification to see Art Nouveau simply as a decorative style based on the work of the Arts and Crafts Movement, interested in the application of natural forms to decorate all forms of art, design and architecture. Art Nouveau should actually be seen as the first 'Anti Design' movement. Art Nouveau designers were not interested in replicating the past. They strove to create a totally new and unique style for the beginning of the twentieth century. The style was created through a wide and diverse range of social, technological, cultural, historical and scientific influences.

Socially the dawn of a new century created an optimistic atmosphere where anything seemed possible, creating an open-minded and sympathetic view of new ideas.

Advancements in materials, such as wrought iron, created opportunities to experiment with new structures, which would have been impossible using traditional materials.

The development of large cities through industrialisation created cultural melting pots. Improved transportation opened up trade links with the Far East bringing new influences from places like Japan. Art Nouveau's stylised organic forms were clearly influenced by Japanese prints with their use of pattern, dynamic lines and flat colour.

Charles Darwin's *Origin of Species* explored the theory surrounding the constant evolution of the natural world. This prompted artists to see natural forms as a starting point for design which they could improve through stylisation and enhancement. This idea of metamorphosis also created the opportunity to combine plant, animal and human forms to achieve an aesthetic look previously never seen. René Lalique, regarded as the master of jewellery design at the time, used nearly every form of Art Nouveau imagery from flowers to graceful women, at times combining a number of forms for his jewellery design.

Although the desire and focus was to create a new modern style, Art Nouveau did not dismiss the past. Celtic and Viking art created an interest in the use of intricate and linear patterns and the Rococo style of the eighteenth century revived the interest and use of delicate curvilinear forms in architecture and interior design.

Typical examples of the Art Nouveau style and influences can be seen in the work of Louis Comfort Tiffany. His famous light designs are clearly influenced by an interest in medieval architecture, with their stained glass windows and the use and stylistic interpretation of animal and natural forms.

'Lotus' lamp by Louis C Tiffany, USA, c.1900

Victor Horta was a pioneer in the use of wrought iron to create flowing organic forms using whiplash curves to increase the dramatic effect of his buildings. He carried the same architectural details into the interior of the buildings to create a feeling of unity.

Hector Guimard (1867–1943) combined stylised plant forms and wrought iron to form structures that seemed organic, where decoration and structure were intertwined and not applied to each other. This made his designs for the Paris metro world famous and an icon of the Art Nouveau style.

The Paris metro 1900. Hector Guimard used natural forms as decoration but combined them with cast iron to form organic looking structures that look like they are growing from the ground.

CHARLES RENNIE MACKINTOSH

Charles Rennie Mackintosh, who was born in Glasgow in 1868, has become one of the most recognisable and popular figures in twentieth-century design. It is impossible to talk about the City of Glasgow without references being made to the School of Art, Willow Tearooms, stylised roses, repeated square motifs and whiplash curves.

As a young man he trained as an architect and attended the Glasgow School of Art where he was influenced by the newly-appointed principal, Francis Newbery.

Newbery encouraged modern thinking, individual creativity and promoted natural forms as a source of inspiration. This, together with the city's prosperity and cultural diversity as a result of its shipping industry, created a new interest in the arts. Mackintosh and Herbert MacNair, together with sisters Margaret and Frances Macdonald, set up 'The Glasgow Four'. Their collaboration on furniture, illustration and interior design proved unsuccessful. Viewed as Art Nouveau it was deemed too decadent and evocative for Victorian Britain. The weird linear work conjured from the female form and ancient Celtic designs created by the Macdonald sisters overshadowed the work of Macintosh and earned the group its nickname 'The Spook School'.

Despite this the group gained widespread recognition for their contribution to modern interior design from the rest of Europe, which had nurtured the extravagancies of Art Nouveau.

However it was Mackintosh's attempts to distance himself from Art Nouveau, promoting function over frivolity, controlled linear geometry over wasteful swirling lines, and also the use of materials for their natural beauty and honesty, that gained Mackintosh individual acclaim from the continent, especially in Germany and Austria.

His focus on function and the honest use of materials in design (which was probably as much to do with economic necessity as any modern philosophy), created his reputation as a modern thinker and pioneer of modern architecture. The commission to design the new Glasgow School of Art in 1896, when he was 28 years old, was constricted by a small budget of £15,000. The rest of his commissions were limited to a small group of friends and patrons. Most of his architectural work was undertaken over a short period between 1896 and 1909.

Mackintosh's work on the School of Art, which took over ten years to complete, created an opportunity for the architect to allow the building to evolve around the human use of space, which was an important aspect of all his architectural work. The evolution and ongoing development of his architecture and design was an approach that frequently led to changes, often leading to spiralling costs and problems with clients.

The Tearooms designed for Miss Cranston show the architect's desire to link the exterior and interior of the buildings to form the unified and complete design concept seen as essential to Mackintosh. They also underline the architect's determination to focus on the buildings' intended use as the main design consideration. The feeling of space was created to ease the busy atmosphere; even his famous high-backed chairs were designed to suit the high hairstyles and hats fashionable with ladies of the period.

Charles Rennie Mackintosh used wrought ironwork extensively as both functional and decorative elements in his buildings. The piece shown here is the exterior signage for the Willow Tearoom Glasgow (1903–1904). The metal used is wrought iron (worked iron), which was named after the process by which it was made. Rods of white hot metal were hammered or rolled together producing an elongated grain structure with a high carbon content. This gave a strong malleable metal which proved resistant to the elements.

Windyhill and Hill House are a homage to the Scottish Baronial style. This national building style in Scotland was viewed as an intrinsic part of the Scottish landscape, providing the architect with a basis for a timeless style admired for its economic use of natural material and monumental status. This can be seen through the use of turreted stairwells and small openings to create windows and doors to form the buildings' facades.

The ladder back chair

Mackintosh strove to design every detail and element for all his architectural projects from exterior to interior light fittings and furniture. He often dismissed the views and wishes of his clients in his attempts to devise a unity in design; this gained him an adverse reputation, and lost him clients.

In 1913 he resigned from his architectural practice and moved to Suffolk, where he intended to be a painter, but he then moved to London and had returned to architecture by 1915. His commissions were limited by the war and he succeeded in completing only a handful of projects. He died in

1928 aged sixty, almost an unknown in the architectural world. It took more than half a century before the genius of Charles Rennie Mackintosh was recognised and celebrated.

The 'Attendant's Chair' by Charles Rennie Mackintosh was designed for the attendant who took the orders from the waitresses in the Willow Tearooms in Glasgow. The lattice back was designed as a divider between the light and dark areas of the restaurant. The chair was a one off and purpose built for these reasons. The illustration shown is a modern replica.

DE STIJL

De Stijl was a style of design that flourished in Holland during the 1920s. It was the vision of the Dutch painter and architect, Theo Van Doesburg, who was committed to creating a 'national style' for Holland, promoting his philosophy of design as editor of *De Stijl*, a magazine from which the movement took its name.

Contributors to the magazine included such influential figures as Piet Mondrian, J.J.P. Oud and Gerrit Rietveld. Simply translated 'De Stijl' means 'The Style'. This group of like-minded visionaries barely knew one another and were only really linked through the magazine and their belief that the architecture of the past should have no influence on the new modern era. They promoted and strove to produce a unified style that would continue throughout all their work, creating a visual harmony. They did not believe in the need for individual style within the arts. To achieve this they completely dismissed ornamentation of any kind in favour of a rather austere aesthetic based on simple geometry created from horizontal and vertical lines, primary colours, and the neutrals colours, black, white and grey.

As a contemporary of the Bauhaus, De Stijl also believed that design should embrace and promote industrialisation and new technologies. They thought that designs should be created from a series of component parts that can be viewed individually or as a complete work.

The rules set out by De Stijl were rarely followed and caused great arguments amongst the group members. This resulted in Mondrian leaving over the inclusion of diagonal lines in the aesthetic style.

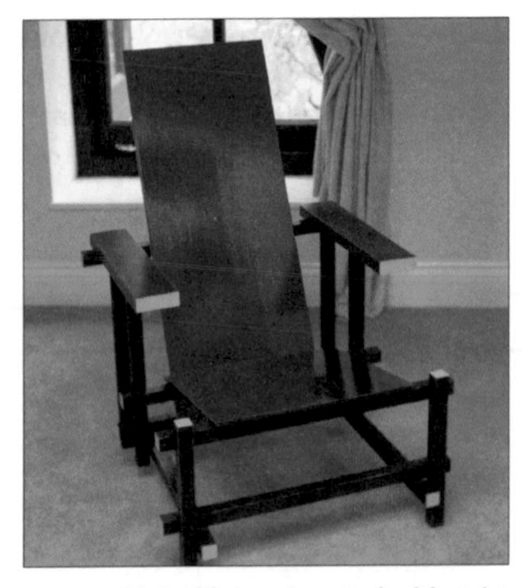

Rietveld experimented with strict geometric structures in his furniture design. The beechwood used in this chair is precut in standard dimensions.

Apart from Mondrian's paintings the design work of Gerrit Rietveld is probably the best example of the De Stijl style. His now iconic 'red-blue chair' is constructed from standard sections of timber arranged in horizontal and vertical lines which combine to form a simple aesthetically pleasing product. His design of The Schroder House also promoted the ideals of the De Stijl movement. It was constructed using modern building materials and allowed for an open-plan structure with few load-bearing walls that produced a flexible living environment that could change to suit an individual's needs. It was created from large intersecting surfaces that were highlighted through the use of primary coloured windows and railings. The building is asymmetrical with large window openings to highlight the building's seemingly simple construction. The house's design and construction was heavily influenced by prefabrication and standardisation. Its simple looking construction suggests elements could be repositioned to form other houses economically with a clear standardised style.

The Schroder House by Gerrit Rietveld

BAUHAUS

The Bauhaus was a school of art and design set up by the architect Walter Gropius in Weimar, Germany in 1919. It aimed to replace traditional education in the arts by breaking down the barriers between fine art and the applied arts. This approach gave hands-on experience in a wide range of disciplines in an attempt to create designers for the new industrial age of manufacture.

Unlike the rest of Europe and America, Germany was in social and economic decline after the First World War. This was reflected in the teachings and philosophies at the Bauhaus which viewed art, architecture and design as a means to improve social conditions by providing affordable housing and products for the working classes.

Its desire to unite art and industry created a new model for teaching that has since been adopted by art and design schools throughout the world. Students followed a foundation course in their first year, studying a wide range of disciplines. They then specialised in a particular field of art or design. To ensure that students were prepared for and could contribute to industry the school was divided into workshops. Each shop had equal input from an artist acting as 'master of form' together with a 'master of applied arts'.

The Bauhaus viewed art and design from a scientific perspective. They introduced the study of economics, sociology and psychology and strove to create rules to govern the use of colour, shape and form. They saw the machine age with its new technology and materials as an economic means to improving the standard of living for the working classes.

Chess set

Geometric forms were favoured by students of the Bauhaus for their simplicity. They believed these forms would lend themselves to industrial manufacture. Colour was limited to primary colours, together with the neutrals – black, white and grey. These were used to enhance the properties of the geometric shapes characterised by the frequent use of the yellow triangle, red square and blue circle – yellow to exaggerate the dynamic qualities of the triangle and blue to accentuate the cool serenity of the circle.

Ornamentation was omitted, again in an attempt to aid mass manufacture and emphasise the functional importance of design over its decorative qualities which were seen as wasteful and unnecessary. 'Form follows function' is a phrase that symbolises the philosophy of the Bauhaus, suggesting that function is more important to good design than form. However, the pursuit of functional excellence would result in a product inheriting an aesthetic quality that would be pleasing to the consumer.

Chair by Marcel Bruer

In reality, the Bauhaus failed to achieve the goal of providing cheap mass-produced products to improve the lives of the working classes. Instead they produced prototypes that were suitable for mass manufacture using modern materials, but were in fact labour intensive products made in the craft room, bought by the middle classes on the grounds of good taste. However, they did pave the way and took an active part in the true machine age that was to take place throughout the postwar period of the 1950s and 1960s.

The Bauhaus Building in Dessau by Walter Gropius

ART DECO

The emergence of Art Deco can be traced back to 1910, but its influence in the world of design was not widely recognised until the 1920s because of the First World War. The term Art Deco was taken from the 1925 Exposition International des Arts Décoratifs et Industriels Modernes, which brought the style to the attention of the public.

The style evolved as both a reaction against Art Nouveau and as a celebration of surviving the restrictions, depravity and poverty of the First World War. It strove to replace the heavy complex floral themes with simple geometric shapes, clean lines and bold relevance to ancient civilizations. It also embraced the modern 'machine age', using the functional forms of airplane wings, ocean liners and car engines as sources of inspiration for decorative design.

Wooden Art Deco Radio

The influence of Ancient Egypt, through the highly publicised discovery of Tutankhamun's tomb and Aztec temples can clearly be seen in radios, jewellery and mantle clocks of the period.

The positive attitude taken by the designers towards the machine age made Art Deco the first true modern style of the twentieth century. It was to have an overwhelming effect on European and American lifestyles. Its distinctive style was applied to everything from cinemas to cigarette holders.

Its popularity and widespread use can be attributed to a number of factors. Its emphasis on the exotic, luxurious and expensive appealed to the nouveau riche of the period. The indulgence and excess of the time created what has become know as 'high' Art Deco, typified by the work of Emile-Jacques Ruhlmann and René Lalique. This faction of Art Deco used extremely expensive materials, such as mother of pearl, snakeskin, sharkskin and bronze inlays to decorate the surface of their designs and used labour-intensive methods of production which created an exclusive fashion for a minority.

However Europe and America were at this time waking up to a new era of fashion and advertising. Hollywood and the film industry brought fashion and glamour to everyone through the popularity of cinema. This created a new consumer who was aware of and demanded fashionable items, at affordable prices. In order to sustain this new customer demand designers could no longer afford to make their own designs. Mass-manufacturing techniques were used on new materials such as bakelite, aluminium, chromed steel and plastic. Designers instructed specialised tradesmen to realise the designs in the Art Deco style.

Art Deco styling was also used to package the new electrical products. The 'sun ray', zigzag, octagon, hexagon and plastic inlays imitating ivory and tortoise shell were applied to household products, such as radios, fridges, toasters and washing machines. Commercial product design was born and was being used to expand and compete in a global market. Products were no longer being bought for their functional application, style was now an important factor with which to entice the consumer.

The Chrysler Building in New York, by William Van Allen

1950s AND 1960s STYLE

After the Second World War, the arts were held in limbo while Europe and America set about rebuilding their societies and cities. However, design was to play a significant role in the recovery of post war Europe and America. The rebuilding programme was on an unparalleled scale and paved the way for a true machine age. Industry and design combined to improve the lives of ordinary people. This collaboration was created through necessity rather than the idealistic dreams put forward by groups such as the Bauhaus or De Stijl. Heavy industrialisation using new materials and manufacturing techniques offered the only efficient way to rebuild after the war. New materials such as plastic, fibreglass and stainless steel were quickly utilised to supply the demand for consumer products and technical goods. A stabilised economy and freedom from rationing at the beginning of the 1950s produced new found wealth for the working and middle classes, creating a demand and appetite for a new and radical style.

The 1950s and 1960s saw social, economic, scientific and technological advancements that created an air of optimism that fuelled and inspired the artists and designers of the day. Television and advertising played a dominant role in creating a lifestyle image that required status symbols such as cars and built-in kitchens with all the labour-saving technology for modern living. This new modern age saw the reintroduction of influential members of the Bauhaus who now had the materials and technology to realise some of their former ideals. Mies van der Rohe had designed steel and glass office blocks as far back as 1921 but only now was he able to build them. Examples include the Seagram Building in New York.

The Seagram Building in New York by Mies van der Rohe

This type of monolithic architecture of the early 1950s created a backlash that prompted designers such as Charles Eams and Eero Saarinen to experiment with a more organic free flowing style of design that eventually gave way to a revival of Art Nouveau during the 1960s.

Rock and Roll created a youth culture for the first time. Young people wanted a style of their own, influenced by the new music and film stars of the period. Fashion changed at a faster pace and a greater demand for new types of products emerged.

Tailfin of a 1958 Cadillac

The use of the molecular structure as a design influence, especially throughout the 1951 Festival of Britain, was an acknowledgement of the pace of scientific progress and its perceived importance to the modern world. Space exploration also played a large part in the creation of the modern style. The launch of 'sputnik' by the Russians spawned an interest in the future as perceived through the eyes of science fiction. Over the top streamlining applied to all manner of products. American cars of the 1950s can be seen to have been especially influenced by the space race.

America became a symbol for the new world, through music, films and economic prosperity. It was viewed as the glamorous place to be. They appeared to build everything bigger, brighter and better than anywhere else at the time and so set the trends and fashion of the period. The emergence of a 'popular culture' centred around the youth of the period, which was larger in number and had more money and more leisure time than ever before. This paved the way during the early 1960s for Pop Art, another influential movement of the period that was widely used in advertising, interior and fashion.

Fashion was now changing at a greater rate and the consumer society of the 1950s soon gave way to the throw-away society. Competition grew as more and more companies evolved to supply this growing market. Industrial design was now a major factor influencing the success of a company.

The pace of change coupled with the diverse range of inspirations created an exciting variety of styles throughout this period from geometric, large-scale buildings to science fiction based clothing and psychedelic record sleeves.

POSTMODERNISM

Postmodernism is a term closely associated with design in the 1970s and 1980s. As the term suggests it is a style that was established after Modernism. Postmodernism is a style of design that was in many ways the opposite of the three main principles of Modernism: form follows function, economy of form and truth to materials. The term postmodernism is now often used to describe anything new, unusual or novel. One of the best-known examples of postmodern design in the 1980s is the Italian group Memphis.

The Memphis style was created by a group of young designers and architects around the catalyst of one of Italy's most successful designers, Ettore Sottsass. Their aim was to reintroduce colour and decoration to design in an attempt to give everyday products a new lease of life, celebrating their form rather than concealing it. This was a complete u-turn for designers who had spent almost fifty years eroding decoration to produce sleek functional products for mass appeal and consumption. Memphis wanted to regain the artistic and creative freedom to create products through individual inspiration. Function was to give way to freedom of expression and intuition.

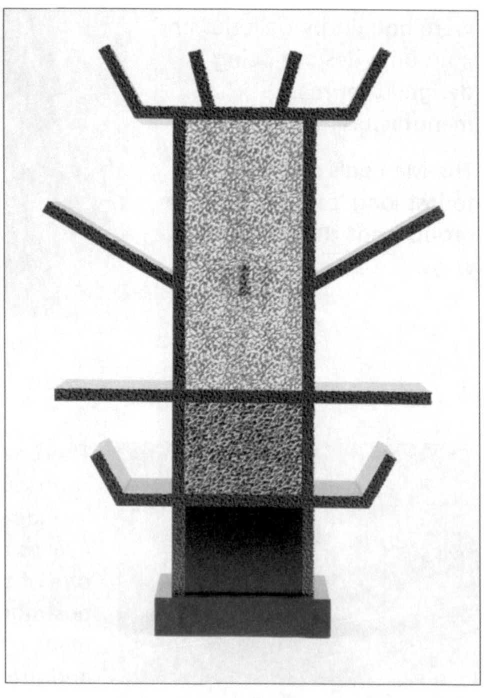

The 'Casablanca' unit by Ettore Sotsass for the Memphis collection in 1981. Small boards are covered in laminated plastic which were put together like a child's game.

Their inspiration came from modern media, television, punk music and comic strips, often creating designs that were perceived as tacky or in bad taste. Their design style was characterised by bright colours and heavy application of surface decoration usually created from laminated plastics, traditionally used in kitchens and bathrooms. Free forms that detracted from the functional expectations of the products, combined with dramatic contrast in material, form and colour, were intended to create a shock factor that forced the viewer to sit up and take notice.

Memphis was an instant success following their first exhibition in Milan in 1981 and put this Italian city at the centre of avant-garde design. The ambition to create everyday products that should be viewed as works of art was established and many of their designs were bought by galleries and museums, despite being designed for mass manufacture.

The Memphis style was not to last long, and in 1988 the group went their separate ways.

Kettle by Michael Graves

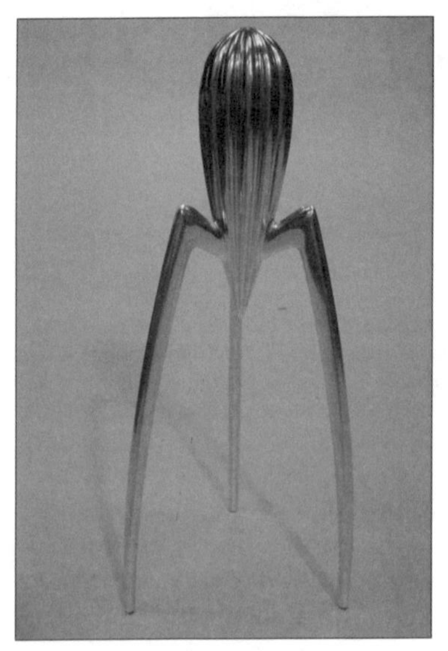

Juicy Salif by Phillipe Starck

There were many other important contributions to this postmodern movement including this kettle designed by Michael Graves for Alessi (1985). It was to become one of the best-known objects of the postmodern period. Philippe Starck also made a notable impact with his streamlined and organically formed products ranging from door handles and toothbrushes to items of furniture. The Juicy Salif lemon squeezer (1990) shown has become a cult item.

3 Design Assignment

- ***Introduction***
- ***Contents of the design assignment***
- ***Design situation***
- ***Research***
- ***Choosing the design task***
- ***What are you expected to produce?***
- ***Section 1: Generating Initial Ideas***
- ***Section 2: Design development***
- ***Good practice exemplars***

INTRODUCTION

Your design assignment is externally set and marked by the SQA. It is worth a total of 70 marks and it will contribute towards 50% of your final grade. The work produced for the assessment must be completed under supervision and remain within your school or centre. However, there is no limit to the amount of preparation and practice that can and should be undertaken before and during this assignment. It is rather like an exam that lasts for several weeks but you have been given the question on the first day. Who wouldn't do extra revision and preparation in these circumstances?

Your design assignment which will be submitted to the SQA will be a maximum of 8 sides of A3 paper. The two exemplars that are shown is this chapter are also available to view and download at: **www.leckieandleckie.co.uk** in the Learning Lab section of the website.

CONTENTS OF THE DESIGN ASSIGNMENT

- Design situation.
- Basic research information, such as lifestyle board/mood board, some ergonomic/anthropometric data, various important sizes and dimensions.
- A choice of four design briefs complete with specifications.

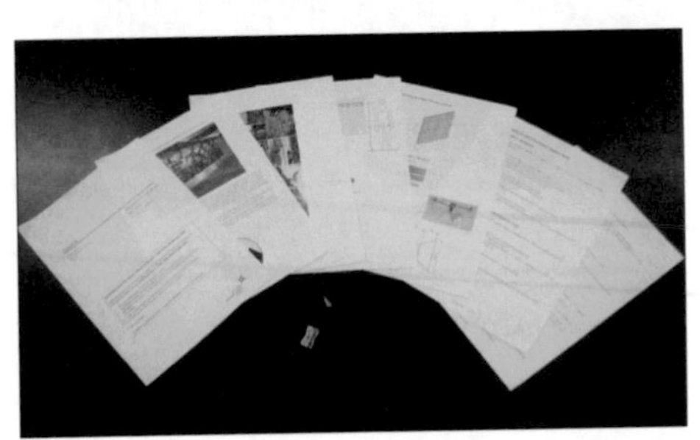

2005 assignment papers

DESIGN SITUATION

The design assignment will usually begin with something called a design situation. This is used to set the scene, put the design in context and add a sense of realism to the assignment. It is all too tempting to skim over this first part or even ignore it altogether. Don't. It's there for a reason. It contains a lot of valuable information that will affect your design – from initial ideas, through its development, to the presentation of your final solution. Close examination will highlight issues and factors such as environmental considerations, user requirements, aesthetic direction and manufacturing constraints which, if carefully analysed, will make idea generation easier, development more productive and justification more relevant.

Action required

Read through the situation and then reread it, this time identifying all the words or phrases that you think could influence your design in some way. However, thinking is not enough. You need to formalise your thoughts. Note down why you consider these words or phrases to be important, what factors will be affected and how this might influence your design. This will stop your designs from becoming too general and superficial. It will also create more areas for development and create better reasoning and justifications.

This activity is very similar to the work done in unit (a) outcome 2: design brief analysis. This is the type of activity that could be done at home in preparation for presenting and developing your initial ideas. Think, plan and research at home, and use the supervised time in class to present your ideas.

Situation

The picture below shows the boot display for a national chain of outdoor stores.

The stores have recently been given a "facelift" and a new rugged outdoor concept is being created. The boot display shelves are mounted on perforated steel sheet which is set into a breeze block wall as shown above. The flooring has a very rough texture. The target market for the stores is 20 to 30 year olds who enjoy outdoor activities. The chain has one hundred stores situated across the UK. The stores are all large and are situated on high street sites.

A number of new shop fittings are required to fit in with the new look. The mood board supplied gives a feel for the type of image and the market that the store is targetting.

Your task is to produce a design proposal for one of the four design specifications provided. Your design proposal must meet all the requirements of the specification and should make use of the research material.

Sample design situation

Here is a design situation taken from the design assignment illustrated on page 34, in which you are asked to design one of the following:

- Seating
  Seating is required which will allow customers to sit and try on boots.
- Sunglasses display unit
  The stores also sell sunglasses and require units to display them.
- Lighting
  Lighting is required to illuminate the boots on the wall display.
- Boot shelf unit.
  The store wants a different shelf unit to feature selected boots.

First, look at the photograph of the shop interior. This will generate some ideas about the aesthetic qualities expected from each of the design proposals. This information will then direct the aesthetics of each of the design tasks. A simple analysis of the picture should come up with words such as 'rough', 'rugged', 'destructive', 'basic', 'simple', 'industrial', which will influence both the initial ideas generation and any further development. Decisions on shape, form, texture and choice of material will all be influenced by this information.

Next, analyse the accompanying text. For example:

- The phrase 'rugged outdoor concept' gives further clues to the expected aesthetic appearance of the designs.
- Special attention is drawn to the fixings which will have a direct bearing on factors such as function, aesthetics, construction, and materials used in the generation and development of the shelf unit.
- Rough flooring could influence the stability and manoeuvrability of the seating or sunglasses unit.
- The age and lifestyle of the target market puts more emphasis on the rugged outdoor aesthetic and also highlights a possible need for fashion to be a consideration.
- The fact that the company has 100 stores should influence the manufacture of the designs. Mass production should be ruled out in favour of batch production, which will influence the construction and materials used.
- The fact that there are large stores situated on the high street suggests a high volume of customers, which could influence the durability, flexibility, safety and maintenance of the design proposals.

Remember that you need to read, analyse and note down all this information. You can then use it together with the research to make an informed choice as to which design task to attempt.

It will help to generate relevant ideas and development, and will also help to explain and justify decisions you take throughout the design assignment.

RESEARCH

You are given research information for a reason, so use it. A large part of the assessment depends on how you incorporate this material into your design assignment. You will be supplied with visual information – usually a mood/lifestyle board. This is expected to steer the aesthetic direction of your product. Decisions about aesthetics should be justified by referencing the mood board and/or the design situation.

Research may also be provided relating to ergonomics, usually in the form of ergonomes and/or anthropometric data tables. This should be used and referenced when making decisions about the position, size and scale of your design.

Other forms of research information will relate more specifically to the tasks set but have in the past included information about manufacture, the environment, additional sizes, limiting factors, etc.

Action required

Analyse the mood/lifestyle board. Identify any prominent styles, fashion, materials and colours that are suggested. It is sometimes helpful to list all the words that come to mind when looking at the visual images. These can then be used to influence the appearance of your initial ideas and trigger developmental decisions.

Note down or highlight what anthropometric data is important and how this is likely to affect your design.

Try to identify how the other information will have an impact on your design in terms of its function, aesthetics and manufacture.

Mood board

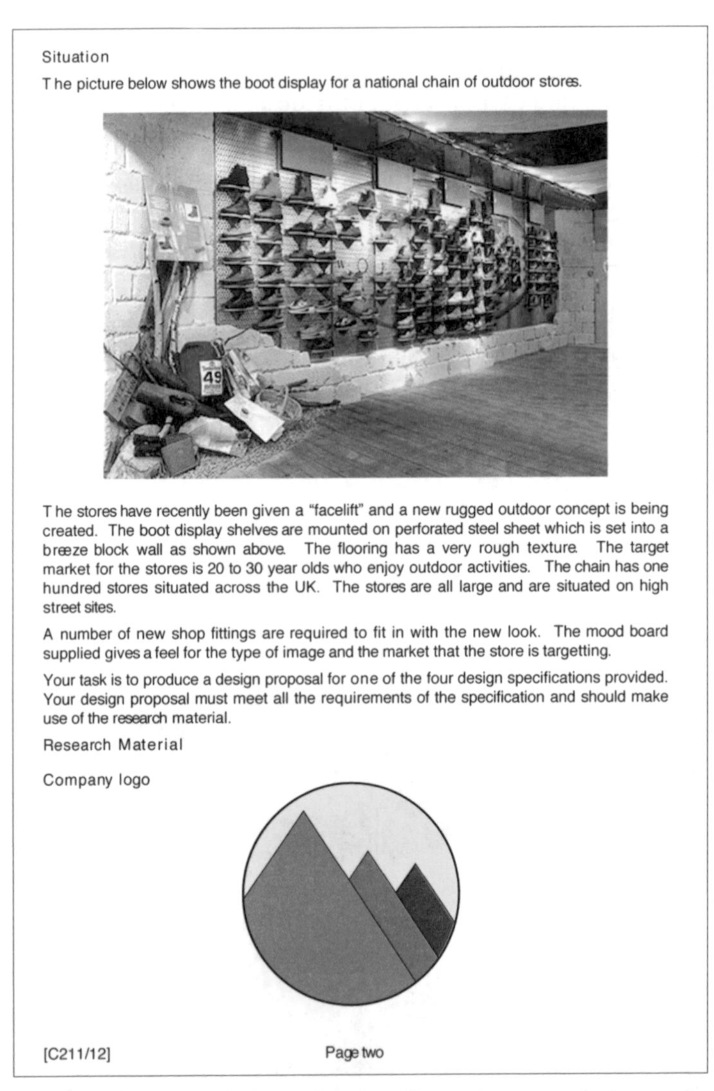

Situation

The picture below shows the boot display for a national chain of outdoor stores.

The stores have recently been given a "facelift" and a new rugged outdoor concept is being created. The boot display shelves are mounted on perforated steel sheet which is set into a breeze block wall as shown above. The flooring has a very rough texture. The target market for the stores is 20 to 30 year olds who enjoy outdoor activities. The chain has one hundred stores situated across the UK. The stores are all large and are situated on high street sites.

A number of new shop fittings are required to fit in with the new look. The mood board supplied gives a feel for the type of image and the market that the store is targetting.

Your task is to produce a design proposal for one of the four design specifications provided. Your design proposal must meet all the requirements of the specification and should make use of the research material.

Research Material

Company logo

[C211/12] Page two

Here is the mood board and research taken from the same design assignment as before. Analysis of the mood board will produce information on the aesthetic direction of the design. You could come up with words such as 'natural', 'extreme', 'dangerous', 'wild', 'rugged', 'open' and 'outdoors', which will influence the choice of materials, colours and textures of the design.

The logo could be used to create a corporate identity which may involve the whole logo, the shape or simply the colours. This could be used on any of the four design tasks.

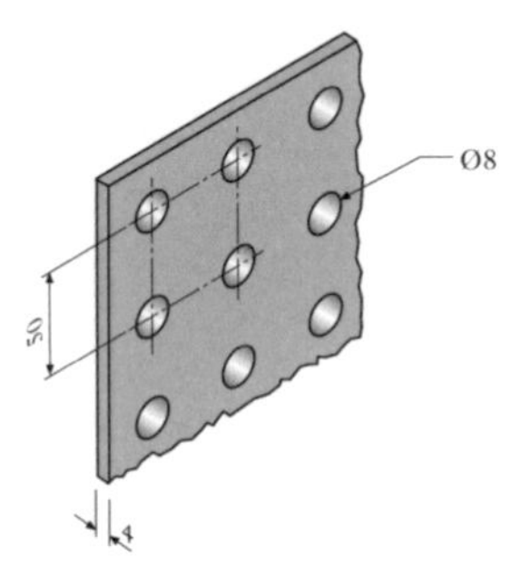

Details and dimensions of the perforated sheet will have an effect on the construction, size and function of the boot display. It will create interesting limitations on the design. Special attention should be given to the diameter of the existing holes, the thickness of the metal sheet and the spacing of the holes.

Typical boot sizes (mm)

Men's size 8
320 L × 120 B × 200 H

Women's size 4
260 L × 100 B × 180 H

Sizes of typical boots will have an influence on spacing and the overall dimensions of the shelf.

Anthropometric data

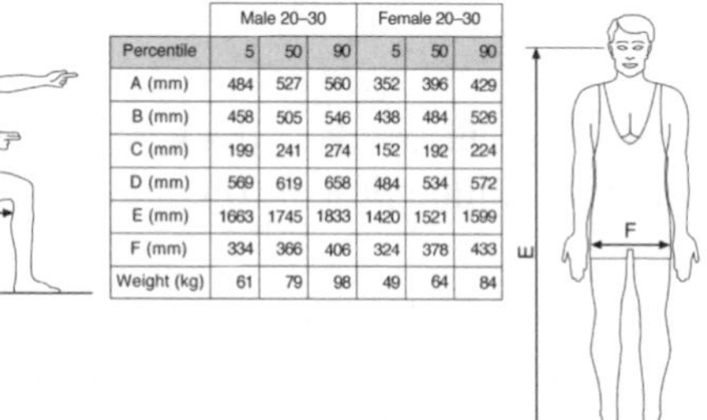

	Male 20–30			Female 20–30		
Percentile	5	50	90	5	50	90
A (mm)	484	527	560	352	396	429
B (mm)	458	505	546	438	484	526
C (mm)	199	241	274	152	192	224
D (mm)	569	619	658	484	534	572
E (mm)	1663	1745	1833	1420	1521	1599
F (mm)	334	366	406	324	378	433
Weight (kg)	61	79	98	49	64	84

Anthropometric data will be important to the development of the seating design and will determine the final size. Careful analysis will highlight which percentile ranges should be used for each dimension.

Example of sunglass sizes (mm)

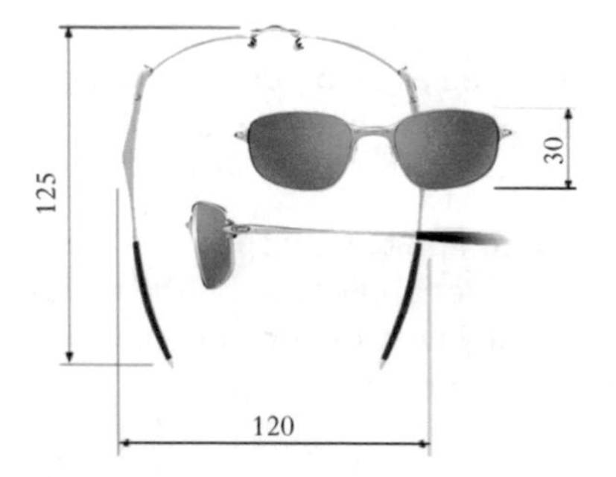

Sizes of sunglasses will influence the development of the design and will determine the final size.

Details of wiring system already in the store

Cables run the length of the shoe display.

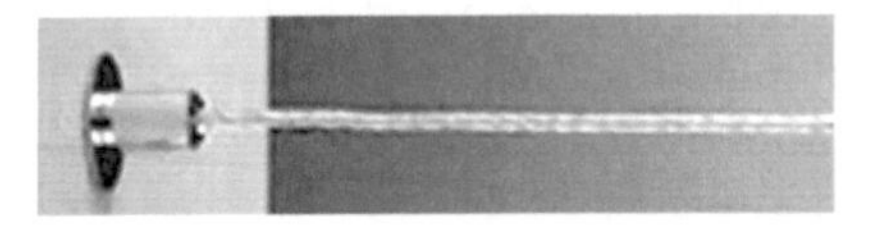

Two existing methods of connecting lights

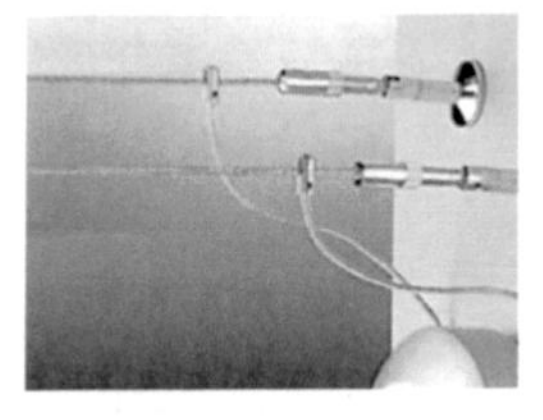

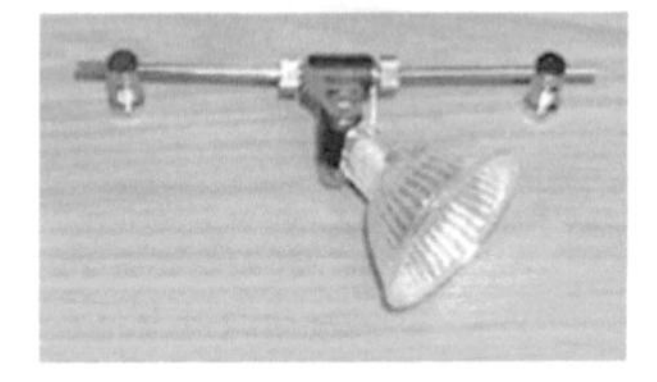

Details of 50 mm halogen bulb

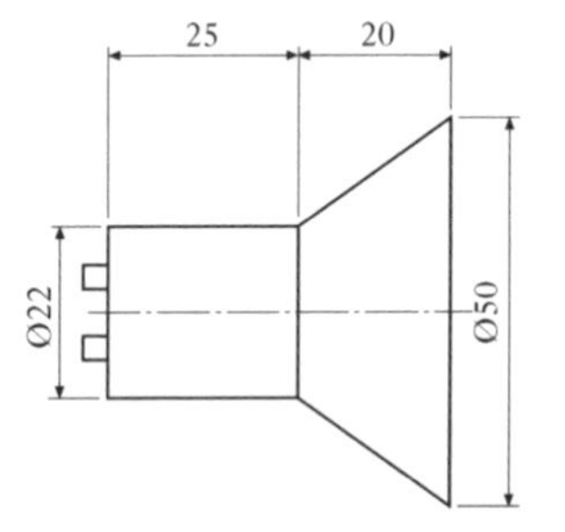

Details of the type of wiring and connection methods of the lights will influence functional and material decisions for the lighting design. The dimensions of the bulb will influence the final size.

Remember that thorough analysis of this research material will make your initial ideas more relevant, highlight potential for development and improvement and provide information for decisions and justifications.

CHOOSING THE DESIGN TASK

Previously this assignment was described as an on-going exam for which you are given the question before you start.You also have the opportunity to choose the question you want to answer.

You will be given four design tasks, complete with specifications, to choose from. Read the four tasks carefully. What appears to be the easiest option does not always offer you the scope for the investigation and development, necessary to complete this assignment successfully.

A good design brief will offer the scope to investigate, explore and develop a range of ideas and solutions through the consideration of function, aesthetics, environment and manufacture. These will all be influenced by the design situation, research material and specifications supplied.

Do not choose the design task because a solution quickly comes to mind. Remember you are being assessed on your application of the design process. Coming up with a solution too quickly removes the need for investigation, creates a reluctance to change and removes the necessity to develop ideas.

Action required

Remember to make an informed choice: choose the task that has a balance of factors which you feel offers the best potential for further development.

Read each task noting down:

- the functions and features suggested by the research material and specification
- important or desirable aesthetic considerations suggested by the research material
- how the environment will affect the function, aesthetics and manufacture of the product
- any constraints created by the specification.

Choose the task that you feel has:

- functional features that are going to be affected by the environment, the user group and its target market, which will require further investigation into areas such as ergonomics, durability and maintenance
- aesthetic direction that may influence the product's shape, size, choice of material, etc, which will in turn create the need for more research and investigation
- environmental considerations that will impact on both the design's function and aesthetics
- manufacturing issues that could be affected by the research and specification supplied.

WHAT ARE YOU EXPECTED TO PRODUCE?

Once you have chosen your task, and before you start the assignment, it is worthwhile outlining what you are being asked to produce, why you are being asked to produce it and how it will be assessed.

It may be worth mentioning at this stage that the design assignment is **not** a measure of talent, flair or aptitude. This would be difficult to assess and rather unfair to the less gifted of us. It is an assessment of our understanding, implementation and skills relating to the design process taught throughout the product design course.

The design assignment is split into three sections. Each section will assess different areas of your knowledge, skill and ability to design products. Marks are awarded for each section.

Section 1

- *Initial ideas* **(15 marks)**

Section 2

- *Development towards a design solution* **(30 marks)**

Section 3

- *Use appropriate communication techniques to convey information regarding the idea generation and development* **(10 marks)**
- *Clearly justify the reasons for decisions taken throughout the folio* **(10 marks)**
- *Effectively communicate the proposed solution* **(5 marks)**

However, to acknowledge and encourage different approaches towards the design process the assignment is marked on a holistic basis. This means that you can gain marks for development from your initial ideas through to your final design solution. Any change at any time that shows consideration will be awarded a mark for development. If, as a result of thorough development, a new concept is introduced that sends your design in a different direction, this would be awarded marks in the initial ideas section. All forms of graphic or written communication are awarded marks throughout the assignment, as are your reasons and justifications at each stage of your work. Although the marks are awarded in three sections, your design work should flow and develop naturally and continuously over the eight A3 pages that you are allowed.

SECTION 1: GENERATING INITIAL IDEAS

Useful skills: research, analysis, evaluation, lateral thinking, quick freehand sketching, three dimensional drawings

Useful knowledge: important design factors, mood boards, technology transfer, materials

The first step to doing well in this section is to have a clear idea of what you are expected to produce. Below is an extract from the SQA Design Assignment Guidance for Higher. It outlines what is required to gain full marks in this section.

Level of response	*marks available*
A wide range of diverse ideas	11–15

- Ideas are diverse and have significant differences between them
- Ideas demonstrate creativity
- Ideas relevant to the specification are produced
- Typically around five alternative ideas are produced
- Alternatively, one concept has been explored by producing a significant range of diverse ideas for major features or components

You are being assessed on your ability to generate a range of diverse ideas that relate to the design task influenced by the additional information provided in the assignment. Simply generating a number of generic ideas may not gain high marks in this section. For example, according to some of the research, the seating design in the sample design assignment would have to have a rugged aesthetic, be suitable for trying on shoes and be movable. Producing different shaped chairs, similar to the ones in your home, would not show understanding. It is vital that you immerse yourself in the problem and carry out thorough research and investigation rather than relying on inspiration and artistry.

Activity – Getting started

Keep your ideas simple to begin with. Consider the function, user aesthetics and the environment when generating your initial ideas. Your analysis will have created lists of important functions and features, possible aesthetic directions, and how the environment affects them. Make sure you use this information in the generation of your ideas. Use annotation to draw attention to how you have used the research to help you generate your ideas. This will ensure they are considered relevant to the design task.

At this stage collecting visual information is essential; it will help generate different ideas and encourage creativity. Undertake further research to find:

- Similar product types that meet the functional requirements or environmental considerations. Adapting an existing design can often be an interesting starting point. However, simply copying a design is unacceptable.
- Information about existing products. This is often a good way to understand specific problems and necessary features of functions concerning the design task. It can often eliminate basic errors quickly. Again, this information should be used to inform and not simply be copied.
- Pictures that could generate an appropriate aesthetic look. This could be more information relating to a particular type of environment, the lifestyle of the users, current fashions and trends or possible themes.

This type of information can be found in books or on the internet and the research can be carried out at home.

Do not rely on inspiration; it can let you down. Practise generating your ideas at home using the information you have gathered. This is preparation for when you have to do it under exam conditions. At home you will be less nervous, more open to ideas and more likely to experiment. The pressure of being in an exam is not conducive to creativity as you will not want to make mistakes. Make the mistakes at home. Go into the exam confident in your ability to present a range of good ideas.

Suitable starting points

Even with all this research and visual information it is sometimes difficult to get started. In an effort to overcome this 'designer's block' try to:

- Identify the major functional elements of the design and produce a basic proposal that fulfils the function, rather like producing a skeleton that could be built upon and developed to meet the other specifications relating to it – aesthetics, environment and manufacture, etc.
- Identify a similar product that fulfils one of the desired outcomes from the design task, such as aesthetics or function, then alter it to meet the other requirements you have identified from your analysis.
- Identify products that produce similar functions to that required and adapt and alter the product to better suit the target market, environment and manufacture.

- Use an image that has the desired aesthetic quality and attempt to add the required functions; adapt it to suit the intended environment; and comment on the use of material or manufacturing considerations.

Instinctive evaluation and development

As soon as you sketch your first idea, irrespective of the quality, you will begin to make judgements and assessments about the design. You will by now be quite conscious of the functional, aesthetic and environmental considerations which are important to your design ideas. At this initial stage your ideas are not expected to be perfect, and highlighting a bad point is a positive step, not a negative one. It is good practice and more natural to sketch out the improved idea beside the original and explain your reasons and thinking behind the improvements.

What is worth annotating?

You will be awarded marks for writing down your justification and reasoning for decisions made. Your comments allow you the opportunity to explain your thinking and display your knowledge of design issues and factors. The examiner is not interested in whether your design is good at this stage. What they will look for is whether or not you have reasons for the decisions and choices you have made.

Comments can be generated from a thorough understanding of the task, the research and the specification. Make sure you make reference to them. Notes could be centred on:

- Reasons for the inclusion of functions and features
- Explanations about how the product will be used
- Important human interactions required and how size, shape and position have been considered
- Why the product will suit the environment
- How the product's shape and choice of materials affect the aesthetics, environment and manufacture.

Try to use phrases or sentences. Use your comments to explain your thoughts. Simply writing words like 'strong', 'interesting' or 'comfortable' beside a design shows little or no understanding.

Appropriate means of communication for initial ideas

- Quick rough sketches mainly done in 2D, or if you are confident, try to produce some 3D sketches as these will communicate your ideas better and offer more potential when it comes to evaluating them.
- At times you might like to show little details that are important to your design; zooming in to sketch these to a different scale works effectively.
- Any medium that can be shaped and moulded quickly will produce adequate block models. Try to select material that will communicate and represent the material you are thinking of using in your proposed solutions.
- Use annotation (notes) to highlight the major elements in your ideas to show that you have analysed and understood the task.

Sample design assignment

Here is an extract from a previous candidate's design assignment. It was awarded high marks in the idea generation section, as each of the ideas presented:

- is clearly different (diverse),
- shows consideration for the design task, makes reference to the specification and has been influenced by the research provided (relevance of ideas).

Each of the ideas is not only visually different but the approach to the design task is also different. This shows a depth of understanding and creativity, providing evidence that the research and specification were used when generating the ideas.

The candidate successfully identified the important functional and aesthetic considerations and incorporated the limiting factor of the existing wall fixings into each of these designs.

Each of the three ideas has been created in an effort to display boots. The shape, size and proportion of each of the designs suggest they are capable of this important function.

Visually, all the ideas would not look out of place in the environment. Words such as simple, modern, industrial and rugged were used when describing the environment. These words could equally be used to describe the designs shown.

Each of the designs has been heavily influenced by the wall fixing that was specified in the design task and detailed in the research provided.

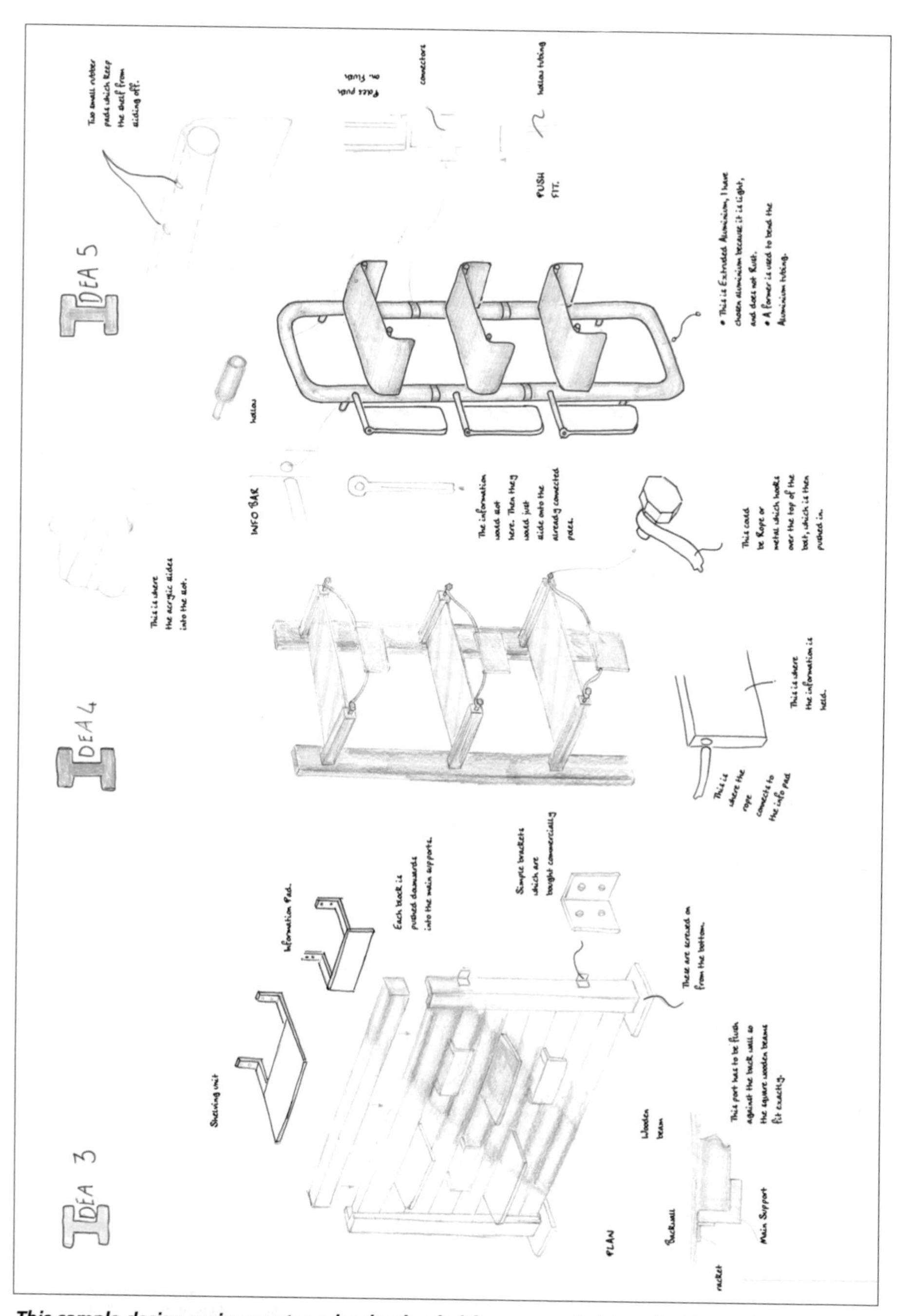

This sample design assignment can be dowloaded from www.leckieandleckie.co.uk

Development

The work on this page gained marks for things other than generating initial ideas.

Each of the initial ideas has been developed in what has been described as intuitive. Some initial sketches highlighted weaknesses when the idea was evaluated against the specification. **Remember constant evaluation creates better design.**

Initially each idea had problems which related to the displaying of information (specification). Each idea was developed to display information better. This only required a little thought and adaptation but it also gained marks for development of the idea.

Other marks for development were awarded for detailing of construction, choice of material and methods of attaching the design to the wall.

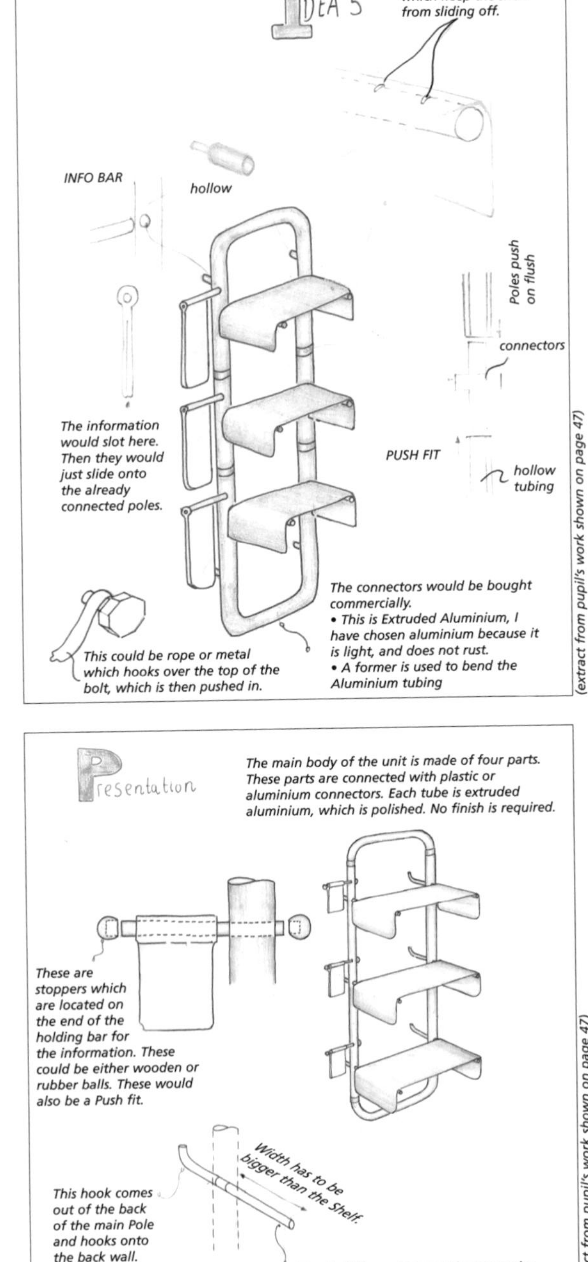

Communication

- Where the candidate felt that their design would not be fully understood they have used annotation to further explain their idea. This gained them marks for communication.
- Annotation has also been used effectively to communicate and highlight material and construction details.
- Sketching techniques have been used appropriately to communicate the candidate's ideas. Three dimensional sketching has been used to create a visual understanding of the ideas.
- Two dimensional sketches have been used to create a better understanding of construction details.
- Scaling up details of the design also helps to describe the idea and create a better understanding.
- The development of each idea has a natural flow. Changes and new ideas seem to suggest themselves naturally.

Recording and justification of decisions taken:

The candidate has made the effort to highlight the most important decisions made during the generation of these ideas, and these all relate to the specification.

SECTION 2: DESIGN DEVELOPMENT

Useful skills: reconciliation of design factors, analysis, evaluation, awareness, appreciation and synthesis

Useful knowledge: relationship of design factors, enhanced knowledge of design factors, deeper understanding of the design process, detailed knowledge of materials, construction and manufacture

As with idea generation, the key to doing well is planning, preparation and research. Don't leave anything to chance.

Level of response	*marks available*
Extensive development of ideas towards a design proposal	25–30

- The design proposal will clearly be derived from initial idea(s) and/or concept(s).
- The most promising idea(s) and/or concept(s) are explored and evolved into a detailed design proposal.
- There is clear evidence of diversity in the exploration of potential solutions.
- There is meaningful reference to the specification at appropriate points.
- Materials and manufacturing processes for the design proposal are clear and justified.
- Clear understanding of design issues is evident.
- The design proposal is clearly evaluated by extensive reference to the specification

Selecting a promising idea

To begin with you will have to identify which idea has the most potential for further development towards a solution. Selecting the idea for development is an important stage, which is often overlooked.

Remember, it will not necessarily be your favourite or the best idea! It is the knowledge and ability required to change, improve, and progress your idea that is being assessed. Choose a design that will make this easy.

A simple evaluation at this stage using the specification as a bench mark is not always informative or useful. It is likely that all the ideas meet the specification as this was a basis for your idea generation in the first place. It may be more informative to ask the questions listed below. This activity will not only highlight the idea with the most potential. It will also focus on development factors and generate the need for more research and experimentation.

Initial evaluation

Try evaluating your design ideas using the following questions. You may need to change, add or ignore some of the focuses depending on the design task.

- **Could the function be improved upon to make it more suited to its purpose?**
  Focus more on what the design is expected to do, who will be using it, problems with its operation, impact of the environment and on its function.
- **Could the aesthetics be improved?**
  Focus more on the influences and impact the environment and user group should have on the overall aesthetic.

- **Could it be easier to use?**
  Focus on who will be using it, how often it will be used, necessary interactions when using, cleaning and maintaining, conditions of use.
- **Could the design present any safety issues?**
  Focus on the potential to cause injury when in use, potential for misuse, legislation and safety standards.
- **Could the design be manufactured?**
  Focus on cost of manufacture, quantity required, suitable manufacturing processes, construction methods, materials.
- **Could the design be improved to better suit its environment?**
  Focus on use within the environment, aesthetics of the environment, position, size and the space available.

A methodical approach

Using these questions together with the specification should enable you to make an informed choice as to which design you want to develop. It should also have highlighted some additional factors and influences that will require consideration, investigation and development. If not, have you really been thorough and rigorous enough.

Having clearly defined areas to work on will make your design development in this section far easier and more productive. It will also create opportunities for preparation and research that can be carried out in advance. It is likely that further research will be required into areas such as function, aesthetics, ergonomics, safety, manufacture, construction, and materials.

As the development progresses the focus will move away from the general function, aesthetic and environmental considerations, and become more concerned with the details relating to issues such as ergonomics, anthropometrics, manufacturing limitations, construction, safety issues, materials, size and scale.

This development section causes the most concern and is often attempted in a rather haphazard manner. Remember we are trying to limit the need for inspiration alone, replacing it with a more reliable methodical approach.

Focus the development around a range of factors relevant to the design task. Use the initial evaluation questions again to highlight areas to develop.

Do not base all your development around one factor, as this will lead to simplistic development that lacks depth and maturity. By considering more factors you will give yourself more opportunities to experiment, explore, improve, justify and reason, all of which will gain you more marks for this section.

Constant evaluation

Good development relies on constant evaluation and re-evaluation of all the changes and decisions you make. You do not have to make constant improvements; many of the changes you make may have little or no effect at all, or they may just prove you were right in the first place. Development can be used to test ideas, prove theories or experiment.

Useful forms of development

Constituent development

Avoid making whole-scale changes to your design idea. This often leads to another concept rather than an improvement or change. Development should be about exploring the possibilities a concept has to offer, not total redesign. Complete change makes evaluation and justification vague and generic and fails to generate focuses for development. Remember the examiner should be able to chart the changes and decisions taken from the initial idea through to the finished proposal. For example, each change in shape, additional feature, or change in material should be explained and justified. If you think of it as a storyboard, explaining the evolution of your design in visual and written media, you will be on the right track.

Focused development

Remember, the examiner needs to know that you have a clear understanding of design issues. To do this you will first have to select relevant issues.

Use the research, analysis and specification, together with your own evaluation, to identify a design issue that you feel could be improved. Use annotations to explain to the examiner why you feel it could be enhanced as this will gain marks for communication (reasons and justification).

Working on one factor at a time will allow you to make better judgements about their effects. Analysing the effect the change to one factor has on the others offers another interesting development opportunity. This is sometimes referred to as the reconciliation of design factors or getting a balanced design.

Ergonomic development

This is a more detailed exploration that will benefit from additional research. Human interaction is always an important aspect of any design and should be easy to understand and incorporate into a design's development. It will have a huge influence on issues such as:

- Function
- Anthropometrics

- Size and scale
- Safety
- Materials

Usually the primary reason for considering ergonomics is to make the product easier or better to use. Make this your goal if you decide to include this form of development.

Reconciliation development

This form of development will show that you have a greater depth of understanding and will lead to enhanced reasoning and justification. It will also generate evidence of diversity and exploration. Design development can often be like a chain reaction. Making a simple change to one element of your design will have an impact on all the other factors. For example, changes to improve the function could involve changing the shape which could in turn affect the aesthetics. This change in shape may influence the choice of material and the method of manufacture, which could influence the construction methods, which could lead to safety issues being exposed. This could lead you to starting the whole process again. This will encourage your design to evolve into a solution and encourage exploration. Typical influence chains may be:

- Function will impact on the aesthetic and ergonomic considerations and have an influence on the choice of materials, construction and manufacture.
- Aesthetics will be determined by the environment or the intended target market and your knowledge of basic aesthetic influences such as proportion, colour and choice of materials.
- Choice of materials will be influenced by the product's function, e.g. strength, rigidity, ductility, weight, durability, hygiene, construction, manufacture.
- Manufacture will be influenced by materials used, quantity required, accuracy.
- Construction will depend on materials required, basic functions such as strength and stability, method of manufacture.

Trial development

Don't be afraid to experiment. Much of the time good design is created from asking 'what if?' rather than stating 'that's it'. Change for change's sake is not necessarily a waste of time. It can often stimulate creative ideas especially if you have run into a mental block. Simply turning an idea on its head (literally) can stimulate an idea. Radical changes usually work best for this activity.

Manufacturing development

This is an essential form of development but should not be the only form of development undertaken. This is often a mistake made by candidates who see this form of development as easy. However, by only considering manufacture the design will lack depth. It will make it difficult to be diverse and explore potential solutions, and it will only demonstrate your understanding of the issues surrounding manufacture.

Manufacturing development is far more technical in its approach than other forms of development and will become more important as your design nears completion. However, don't leave manufacturing considerations until the last page of your development. Good understanding of materials and the manufacturing processes should be evident throughout this assignment from initial ideas onwards. One of the main concerns in any design has to be: how it will be made, if it can be made and what it should be made from. This has a direct influence on how well the product performs, from its function and aesthetics to safety, cost and environmental impact. Make sure you highlight any limiting factors in the research or specification that could influence the design's manufacture, such as the cost of each item or the number required. Remember, some manufacturing techniques will not be suitable.

Working to scale

Beginning to consider and work to a scale usually creates problems for the designer. It should not be considered as the end of the design process. What may have appeared interesting as a small sketch can become an unusable featureless object when drawn to scale. At this time, faults are often exposed that require further development.

An effective and informative method of working to scale is to use models. Whether it is a simple card model of an initial idea or a full sized prototype, models offer great potential for analysis and evaluation, mainly due to their 3D form. They are essential if you want to assess human interactions as they offer the potential to actually interact. They can also be used to test ideas that may only be good in theory. Having to make a model will generate a better understanding and will encourage development. At times it may only be necessary to make a part of the design to scale in order to understand a construction detail or a specific ergonomic interaction. Be sure to record any changes that you make to a model, take lots of photographs and write down what made you change or modify the design. You will forget if you don't write it down. Remember the examiner is interested in your thoughts and opinions.

Scale drawings can also be produced when developing a design but they are of limited use and are best used as part of the presentation of your final proposal.

Appropriate means of communication for development

- More detail in 2D sketching, use of scale to zoom in on details, sectional views could be used to explore and display construction and interior workings.
- Increased use of 3D sketching containing more detail and refinements to explore the form and function with more realism.
- Exploded views that could be used to display information about construction.
- Block and scale models could be used to explore aesthetics and complex 3D forms.
- More detailed full size models or prototypes could be used if the design requires testing or a part of the design may require testing.
- Increased use of colour and rendering could become necessary as the design reaches its completion stages.
- Scale models or prototypes are a useful means to evaluate and develop a design. They are also an excellent method of communicating your design ideas if recorded properly.
- Experiments with anthropometrics will require the use of scale drawings if they are to have any real meaning or value. These do not have to be formal drawings – sketches can be used if they have a scale reference such as an ergonome or photograph.
- Computer generated graphics are also useful at this stage as they can generate more insight into the design and highlight possible problems or areas for further development.

GOOD PRACTICE EXEMPLARS

So far the design assignment has been looked at in separate sections. Extracts have been used to highlight specific points and good practice, but an emphasis has been placed on a more natural and holistic approach throughout the chapter. Following a strict linear approach may appear to be the easiest option, but in fact this stems creativity, reduces the natural flow of ideas and makes development increasingly difficult.

To see how a more honest and natural approach works in practice, another candidate's response to the assignment will be used as an example of good practice.

The candidate also opted for design task 4 – boot shelf unit. This was thought to offer the candidate enough scope to be creative but had limits and constraints that would encourage thorough and detailed development.

The first two pages will examine in detail the candidate's response to section 1, initial ideas, which was awarded 15 marks.

Remember marks are awarded for ideas that are **diverse, creative** and **relevant to the task**. (The candidate's response can be found on pages 59 and 62.)

Diverse

By looking over the candidate's work in this section it is clear that they have produced a number of different ideas. However, this has not been achieved by superficial changes in shape or replicating existing designs. The diversity is found on many levels, made possible through analysis and understanding of the design task. This has allowed the candidate to explore different methods of displaying boots, attaching the shelving to the wall and displaying information, using a variety of materials and construction methods. The different approaches to the functional problems have generated different looking designs that have a sense of reality.

Simply producing a range of different shaped shelves would not have gained many marks.

Creative

Creativity has been demonstrated throughout this section as the candidate has strived to work within the confines of the design task. The analysis of the task and the problems it presented has forced the candidate to create solutions and experiment with ideas. The candidate has worked continuously to produce ideas that reflect the important functions and features, which have also been influenced by the environment and aesthetic direction.

Creativity is evident in the functional aspects of each design. Each idea has provided a different method of displaying the boots, displaying the information and fixing the design to the wall.

The candidate has been creative with her use of materials, exploring potential solutions due to the properties of the material.

Relevant to the task

All the designs presented have been produced in answer to the design task. This can be seen both visually and through detailed annotations. All notes refer in some way to the specification, research or situation.

Aesthetically each of the designs would suit the environment which was described as rough, rugged and industrial. This has influenced the choice of material and functional look of each of the designs. Each of the designs appears to have been created to hold and display both the boots and information. Efforts have been made to incorporate the fixings suggested in the research.

Tips for success

1. **Get the maximum credit for your ideas. Use a range of drawing techniques and annotations to fully explain your thinking.**

   Idea 1 (page 60) clearly shows that the candidate has a thorough understanding of the task. Not only has the boot shelf been created but the candidate has shown that they understand the problems associated with fixing it to the wall, displaying information and incorporating the company logo.

   Focus on the information pad also shows greater insight and has created some natural development.

2. **Present ideas as they occur. The act of sketching an idea often helps you think of other ideas that need further explanation.**

   Idea 2 (page 61) has a range of details that further explain the candidate's thinking. This clearly shows the candidate has a good understanding of the problem they are trying to address.

3. **Use annotation.**

   Much of the thinking behind these designs would have been lost if the candidate had not written down some of her thoughts. The notes that accompany idea 2 show that problems have been thoroughly considered, such as:

   - the shoes slipping off the metal shelf
   - the simplicity of its construction
   - why there are two sizes of shelf.

   All the comments show greater understanding and relevance.

4. **Stick to the task.**

   Make a genuine attempt to produce solutions to the task set. It may appear easier to ignore the specification, but the design will lack relevance, offer no incentive to develop further and be difficult to annotate.

   All of the initial designs have been detailed, developed and annotated in some way through reference to the task, not simply for the sake of it.

5. **Display wider knowledge.**

   Idea 5 (page 71) displays a deeper knowledge and understanding of the product's function and construction through details, development and written text. This shows a greater insight and provides a more mature idea.

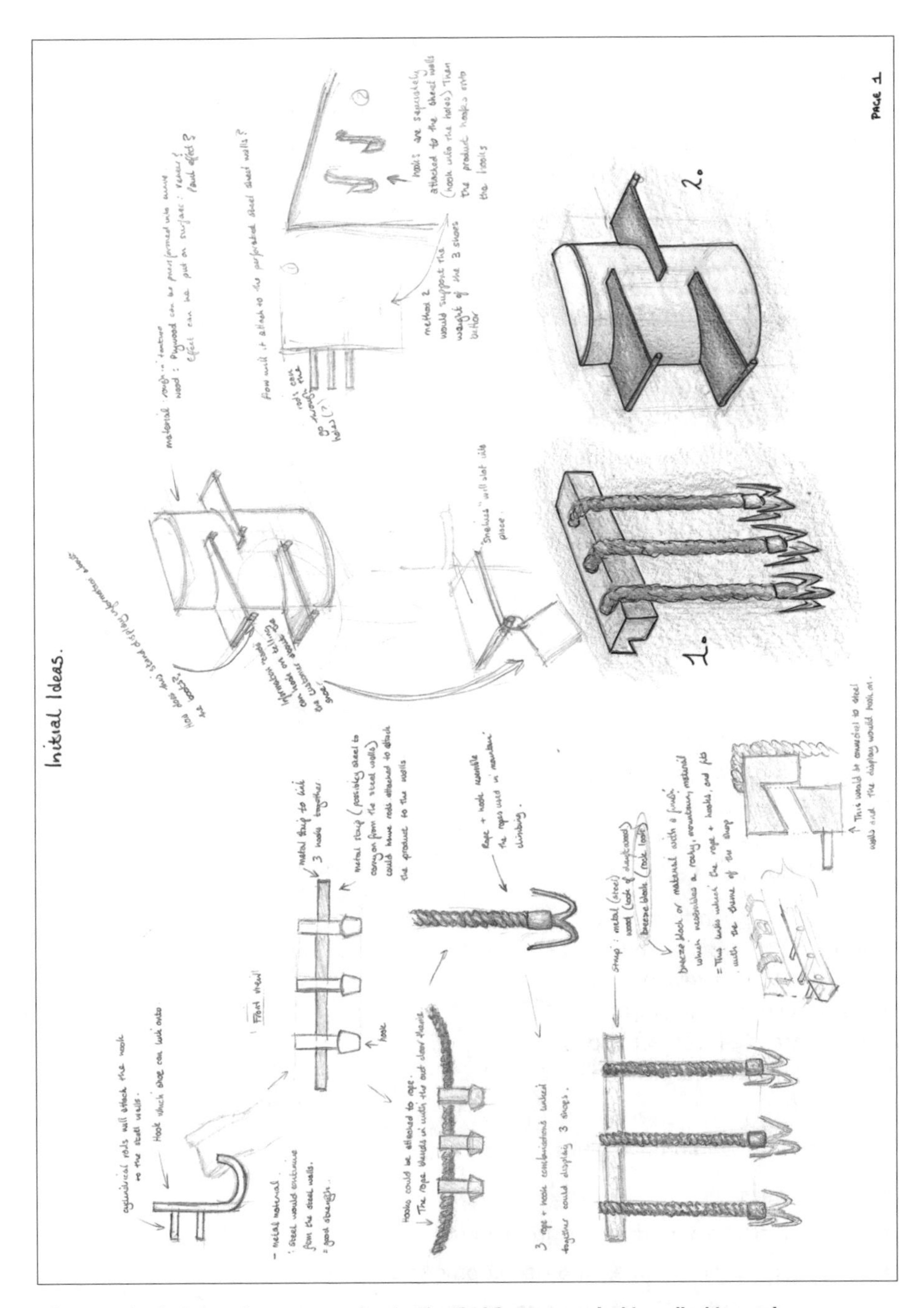

This sample design assignment can be dowloaded from www.leckieandleckie.co.uk

Tip

Present ideas as they occur. The act of sketching often helps you to think of other ideas that require further explanation and development.

A closer look at idea 1 reveals how much thought has gone into its creation. Small changes to the initial idea have been made by the candidate in their attempt to answer the design task more fully.

1 hook changed to 3 to accommodate the number of shoes stipulated in the design task.

Steel changed to rope to better reflect the outdoor theme.

Rope inspired thoughts of climbing resulting in the hooks.

Changes to the material created unforeseen problems for fixing the display to the wall, which the candidate acknowledged and addressed.

This shows the candidate has identified the basic elements important to the design's initial success and attempted to generate ideas relevant to the task, through considered evaluation.

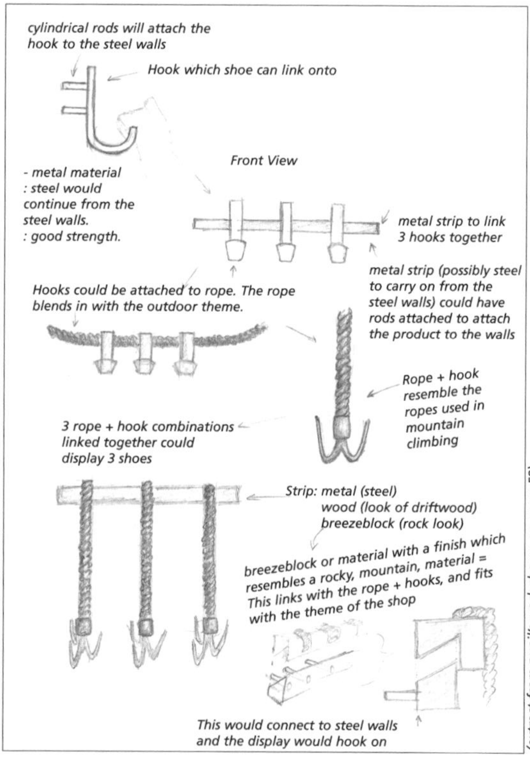

(extract from pupil's work shown on page 59)

The candidate would have gained credit for:

- Initial ideas
- Development towards a design proposal
- Communication of ideas and development
- Recording and justification of decisions taken.

Tip

Make every effort to answer the design task. This should involve constant evaluation of your own ideas. It may appear easier to ignore the task and specification when generating ideas, but in reality it makes creativity and development more difficult.

A closer look at idea 2 reveals the efforts made to produce a relevant solution.

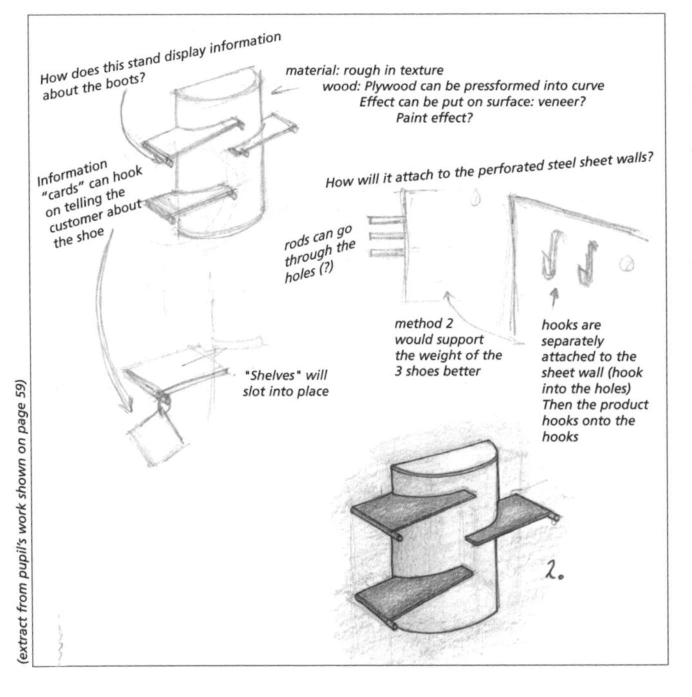

(extract from pupil's work shown on page 59)

Here the candidate has highlighted some of the faults and shortcomings of the initial design.

Efforts have been made to address these problems stimulating meaningful development.

Minor changes have created a more relevant idea.

The design now has the ability to fix to the wall in the prescribed manner and display information.

Here the candidate has displayed a thorough understanding of the design task and specification, demonstrating how it impacts on their initial ideas.

The candidate would have gained credit for:

- Initial ideas
- Development towards a design proposal
- Communication of ideas and development
- Recording and justification of decisions taken.

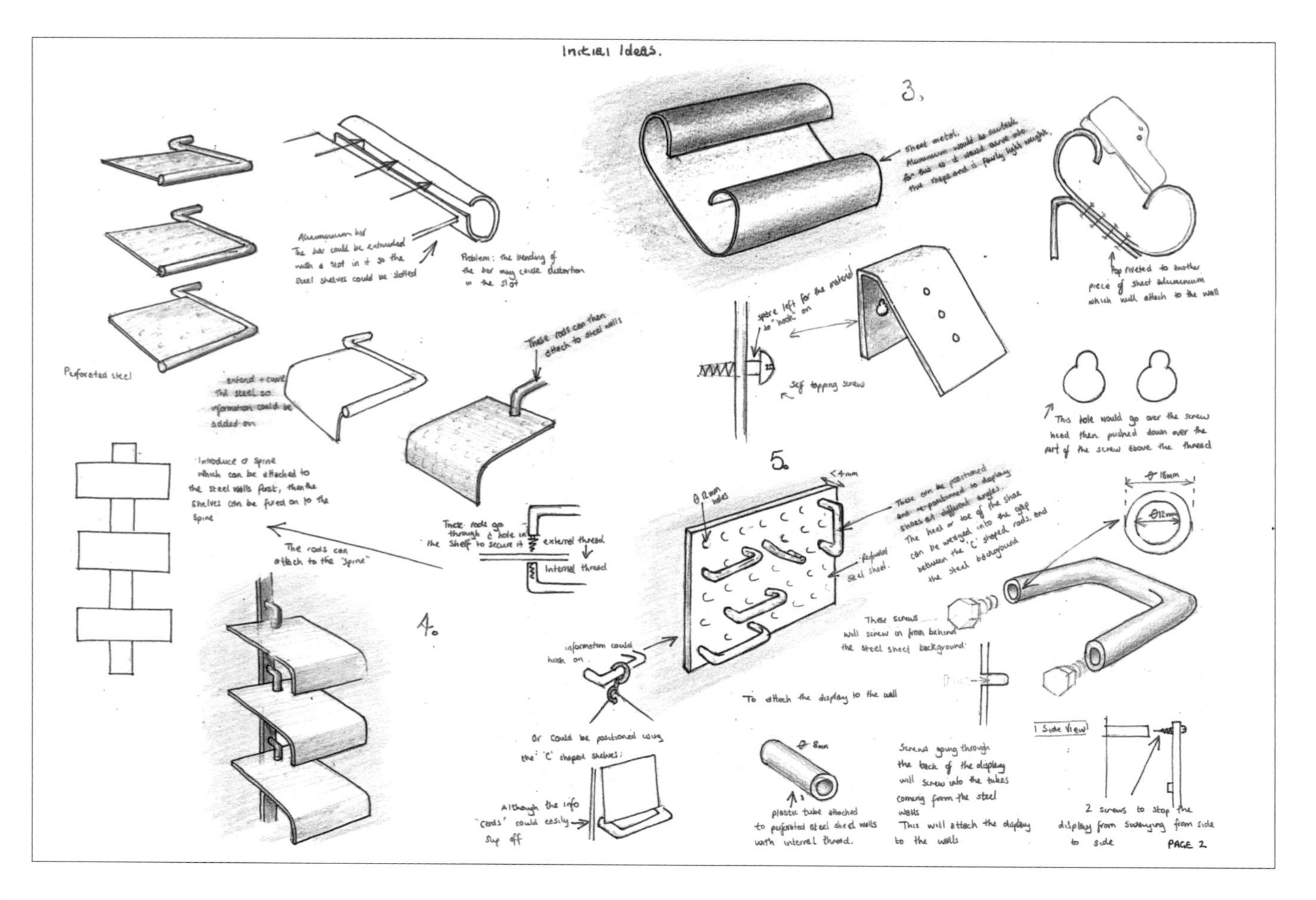
Initial Ideas.
Perforated steel
Aluminium bar
The bar could be extruded with a slot in it so the steel shelves could be slotted
Problem: the bending of the bar may cause distortion in the slot
These rods can then attach to steel walls
Introduce a spine which can be attached to the steel walls first, then the shelves can be fixed on to the spine
The rods can attach to the "spine"
These rods go through a hole in the shelf to secure it
external thread
Internal thread
4.
3.
Self tapping screw
This hole would go over the screw head then pushed down over the part of the screw above the thread
5.
These screws will screw in from behind the steel sheet background.
information could hook on
Or could be positioned using the 'C' shaped shelves:
Although the info 'cards' could easily slip off
To attach the display to the wall
plastic tube attached to perforated steel sheet walls with internal thread.
Screws going through the back of the display will screw into the tubes coming from the steel walls
This will attach the display to the walls
Side View
2 screws to stop the display from swaying from side to side
PAGE 2

Tip

Make sure you get maximum credit for your ideas. One drawing for each of your initial design ideas may not be sufficient to fully explain your thinking. Use a range of drawing techniques and annotations to fully explain your thinking.

Idea 3 viewed on its own is difficult to understand and could have been viewed as a weak solution. When supplemented with additional drawings and annotations, it suddenly becomes a very mature and reasoned response.

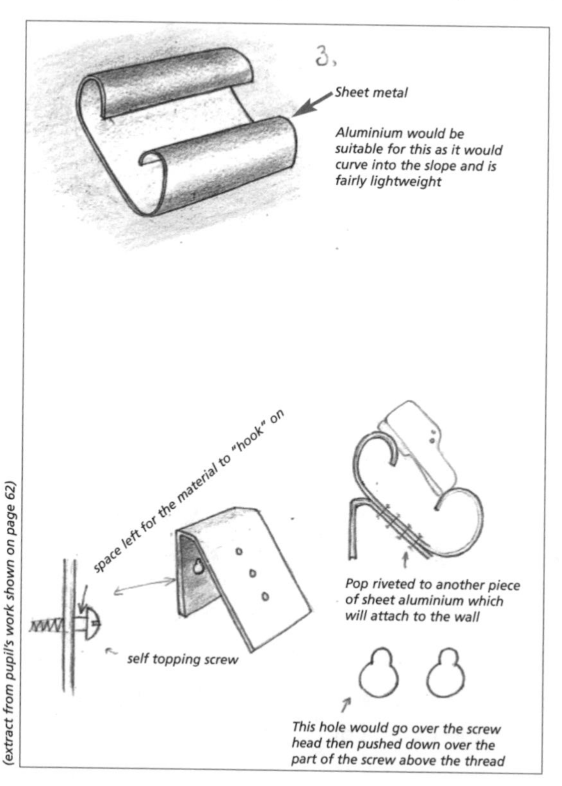

(extract from pupil's work shown on page 62)

The side elevation creates a better understanding of how the design will display the boots.

The 3D detail of the fixing bracket helps to visualise how the display is to be constructed and fitted together.

Further detail has been supplied on the fixing method.

Annotation highlights knowledge of materials and construction methods.

This shows that the candidate has considered methods of communication, identified weaknesses in the presentation and provided additional forms of communication.

The candidate would have gained credit for:

- Initial ideas
- Development towards a design proposal
- Communication of ideas and development
- Recording and justification of decisions taken.

Tip

Use annotations and also highlight comments that show you have a deeper understanding of the problem. This offers a quicker means of communication at this stage of the design process.

Idea 5, which is less visual and more technical, has a large amount of text to supplement it. This focus on the more technical elements may provide information that could be used later in the development stages.

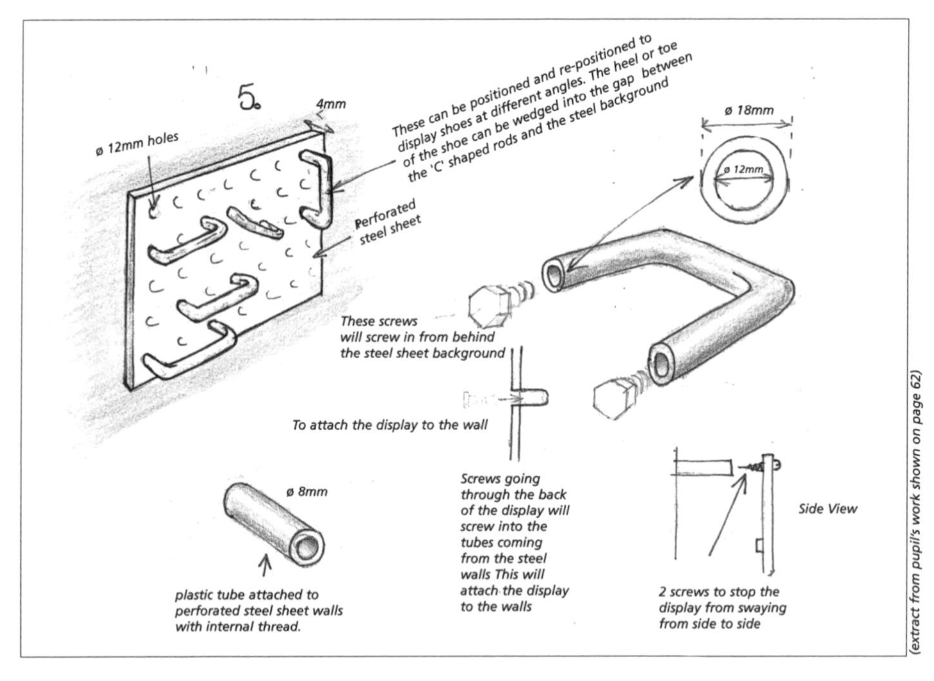

(extract from pupil's work shown on page 62)

This extract shows that the candidate has not only provided ideas that are visually interesting but have been considered in more depth.

The candidate would have gained credit for:

- Initial ideas
- Development towards a design proposal
- Communication of ideas and development
- Recording and justification of decisions taken.

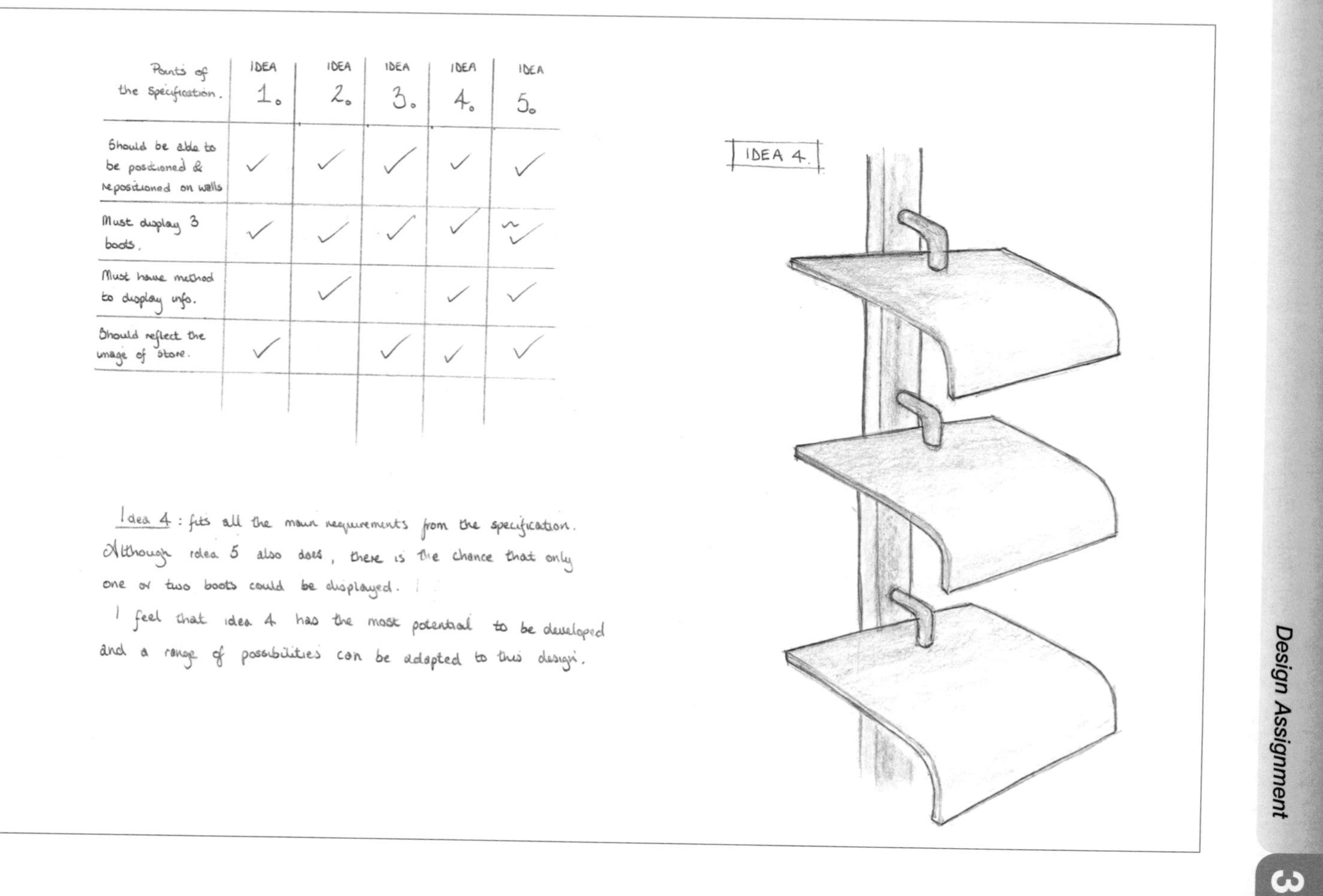

Points of the Specification.	IDEA 1.	IDEA 2.	IDEA 3.	IDEA 4.	IDEA 5.
Should be able to be positioned & repositioned on walls	✓	✓	✓	✓	✓
Must display 3 boots.	✓	✓	✓	✓	~✓
Must have method to display info.		✓		✓	✓
Should reflect the image of store.	✓		✓	✓	✓

Idea 4 : fits all the main requirements from the specification. Although idea 5 also does, there is the chance that only one or two boots could be displayed.

I feel that idea 4 has the most potential to be developed and a range of possibilities can be adapted to this design.

Tip

A formal evaluation is not a necessary requirement for the design assignment, but it offers an opportunity to show that ideas have been referenced to the specification and promising ideas have been identified and justified.

Here the candidate has produced a chart to evaluate each of the 5 initial design ideas, highlighting why idea 4 offered the most potential. Written text also supplements and justifies the reason idea 4 offered potential for development.

Remember, it does not necessarily have to be the best or most complete design that is taken forward, it's the one with the most potential.

Points of the specification.	*IDEA 1.*	*IDEA 2.*	*IDEA 3.*	*IDEA 4.*	*IDEA 5.*
Should be able to be positioned & repositioned on walls	✓	✓	✓	✓	✓
Must display 3 boots.	✓	✓	✓	✓	✓
Must have method to display info.		✓		✓	✓
Should reflect the image of store.	✓		✓	✓	✓

Idea 4: fits all the main requirements from the specification.

Although idea 5 also does, there is a chance that only one or two boots could be displayed.

I feel that idea 4 has the most potential to be developed and a range of possibilities can be adapted to this design.

(extract from pupil's work shown on page 65)

The candidate would have gained credit for:

- Development towards a design proposal
- Communication of ideas and development
- Recording and justification of decisions taken

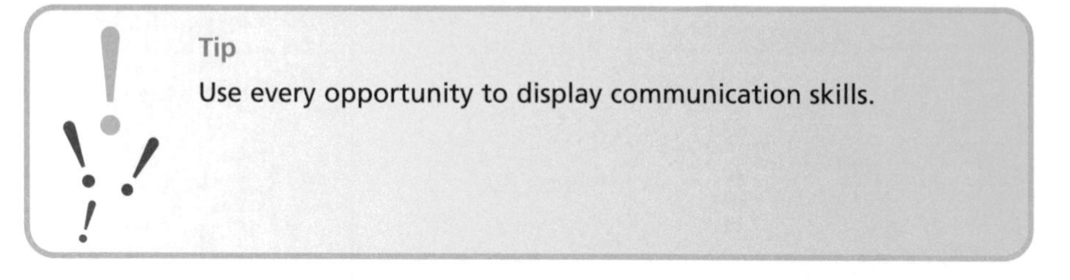

Tip

Use every opportunity to display communication skills.

The chart on its own would not have been effective in identifying the idea with the most potential. The 3D rendered sketch at this point helps to communicate why it offers the most potential. It is easy to imagine how the design would work and how it could be modified.

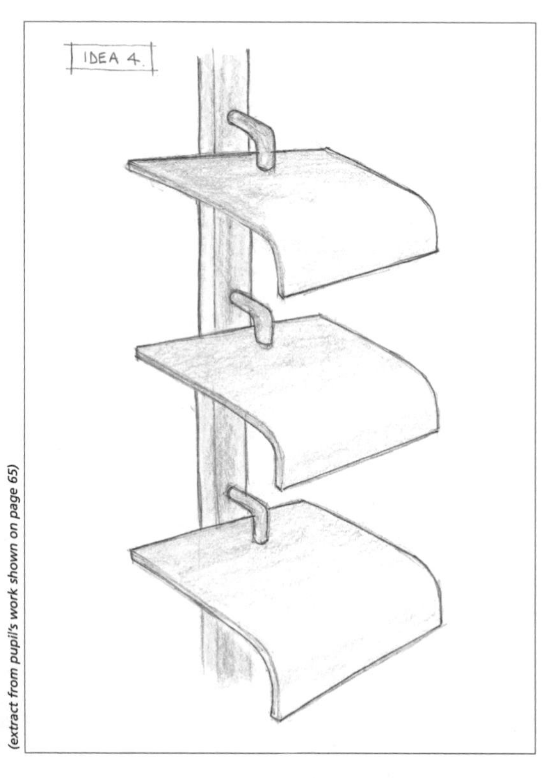

(extract from pupil's work shown on page 65)

The rendered drawing suggests that the candidate has considered the aesthetic qualities as well as the functional properties of the material being used.

The candidate would have gained credit for:

- Development towards a design proposal
- Communication of ideas and development
- Recording and justification of decisions taken.

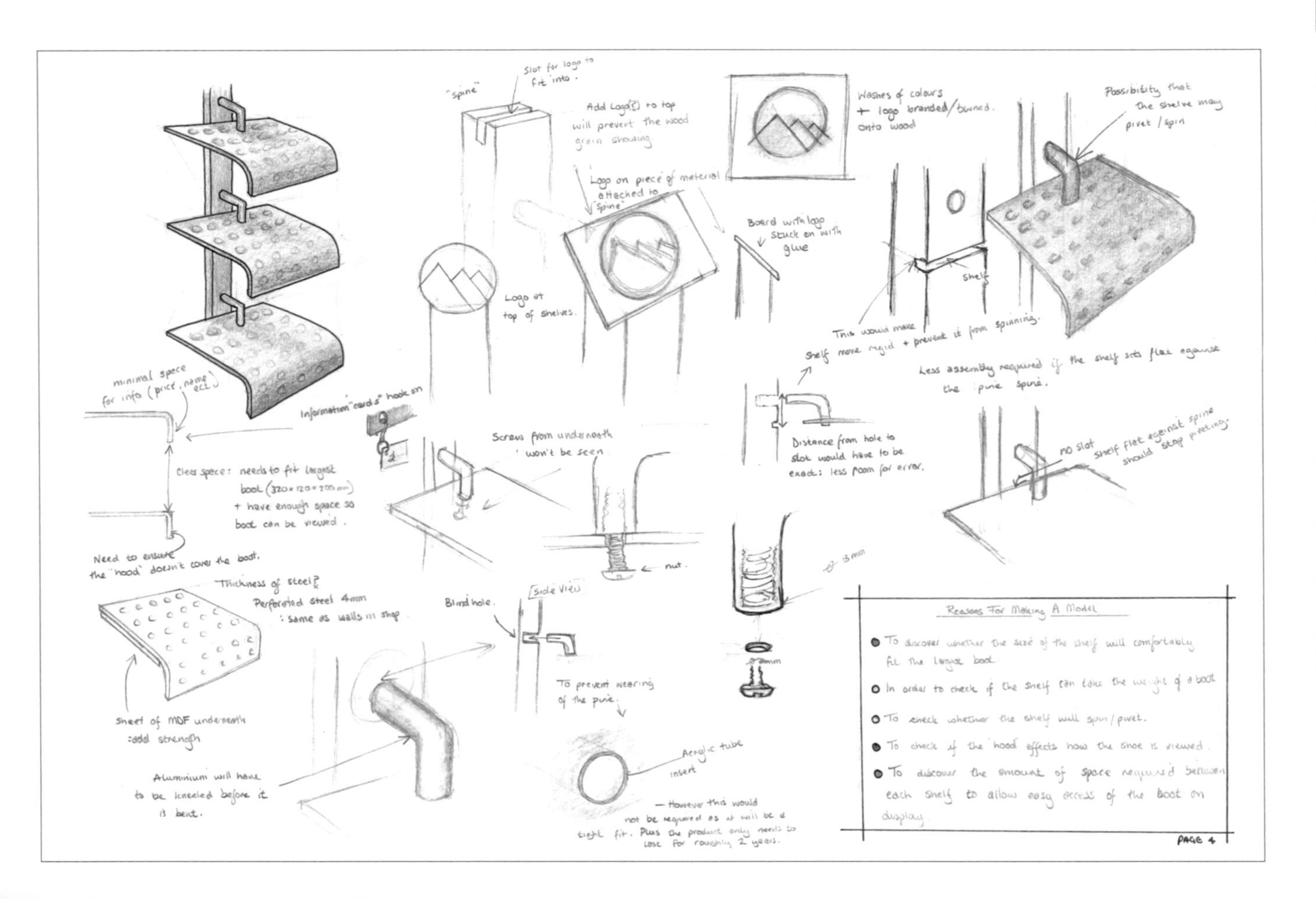
"spine"
Slot for logo to fit into.
Add Logo(?) to top will prevent the wood grain showing
Logo on piece of material attached to 'spine'
Logo at top of shelves.
Washes of colours + logo branded/burned onto wood
Board with logo stuck on with glue
Possibility that the shelve may pivot / spin
Shelf
This would make shelf more rigid + prevent it from spinning.
Less assembly required if the shelf sits flat against the pine spine.
minimal space for info (price, name ect.)
Information "cards" hook on
Clear space: needs to fit largest boot (320 x 120 x 200 mm) + have enough space so boot can be viewed.
Need to ensure the 'hood' doesn't cover the boot.
Screws from underneath won't be seen
nut
Distance from hole to slot would have to be exact: less room for error.
no slot
shelf flat against spine should stop pivoting.
13mm
Thickness of steel?
Perforated steel 4mm
: same as walls in shop
Sheet of MDF underneath : add strength
Blind hole.
Side View
To prevent weering of the pine.
Acrylic tube insert
— However this would not be required as it will be a tight fit. Plus the product only needs to last for roughly 2 years.
Aluminium will have to be kneeled before it is bent.
Reasons For Making A Model
To discover whether the size of the shelf will comfortably fit the largest boot.
In order to check if the shelf can take the weight of a boot
To check whether the shelf will spin / pivot.
To check if the 'hood' effects how the shoe is viewed.
To discover the amount of space required between each shelf to allow easy access of the boot on display.
PAGE 4

Tip

Highlight the comments that are important to the design's development.

In this extract the candidate has used colour pencil to highlight how the research has been considered when developing the design.

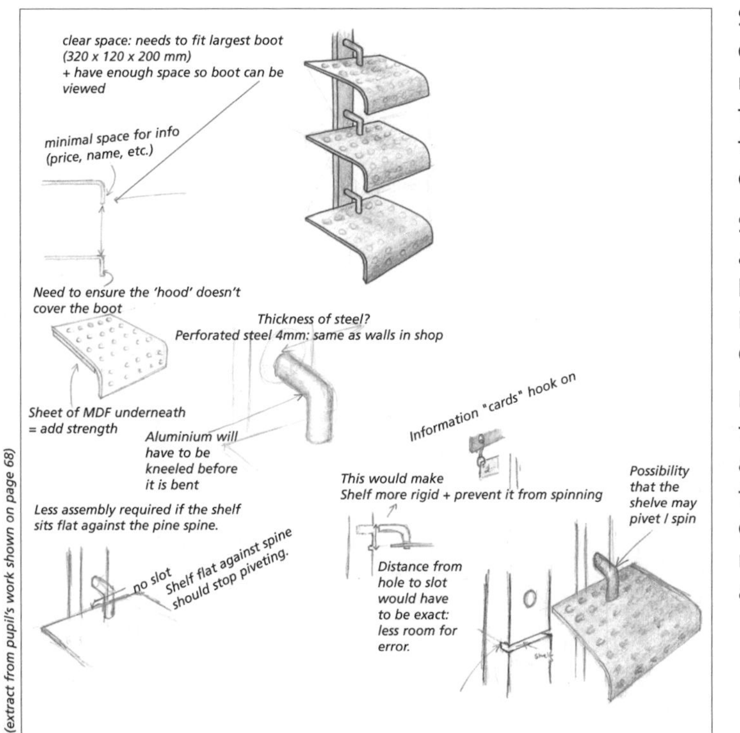

(extract from pupil's work shown on page 68)

Size and dimensions are now beginning to feature in the design's development.

Smaller details are now beginning to influence the design.

Moving away from aesthetics and basic function, construction is now becoming an issue.

This shows that the candidate is beginning to focus their thoughts, moving away from the general concepts towards a detailed solution.

The candidate would have gained credit for:

- Development towards a design proposal
- Communication of ideas and development
- Recording and justification of decisions taken.

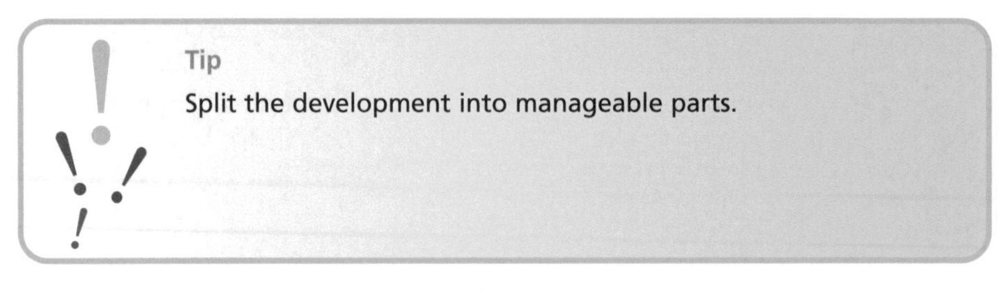

Tip

Split the development into manageable parts.

Rather than trying to improve the whole design at once, the candidate has selected elements that are important and worked on them separately.

Here the candidate identified that the logo had not been used, which was stipulated in the brief. Incorporating it into the design offered opportunity for creativity, development and justification.

Concentrating on small elements of construction allows for greater depth in the development.

This shows that each part of the design has been considered, recorded and developed to a high standard.

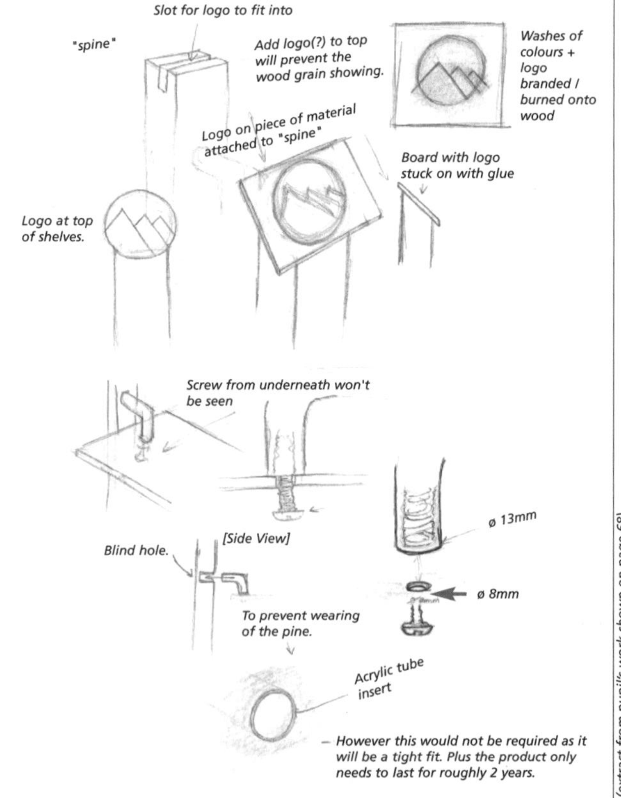

(extract from pupil's work shown on page 68)

The candidate would have gained credit for:

- Development towards a design proposal
- Communication of ideas and development
- Recording and justification of decisions taken.

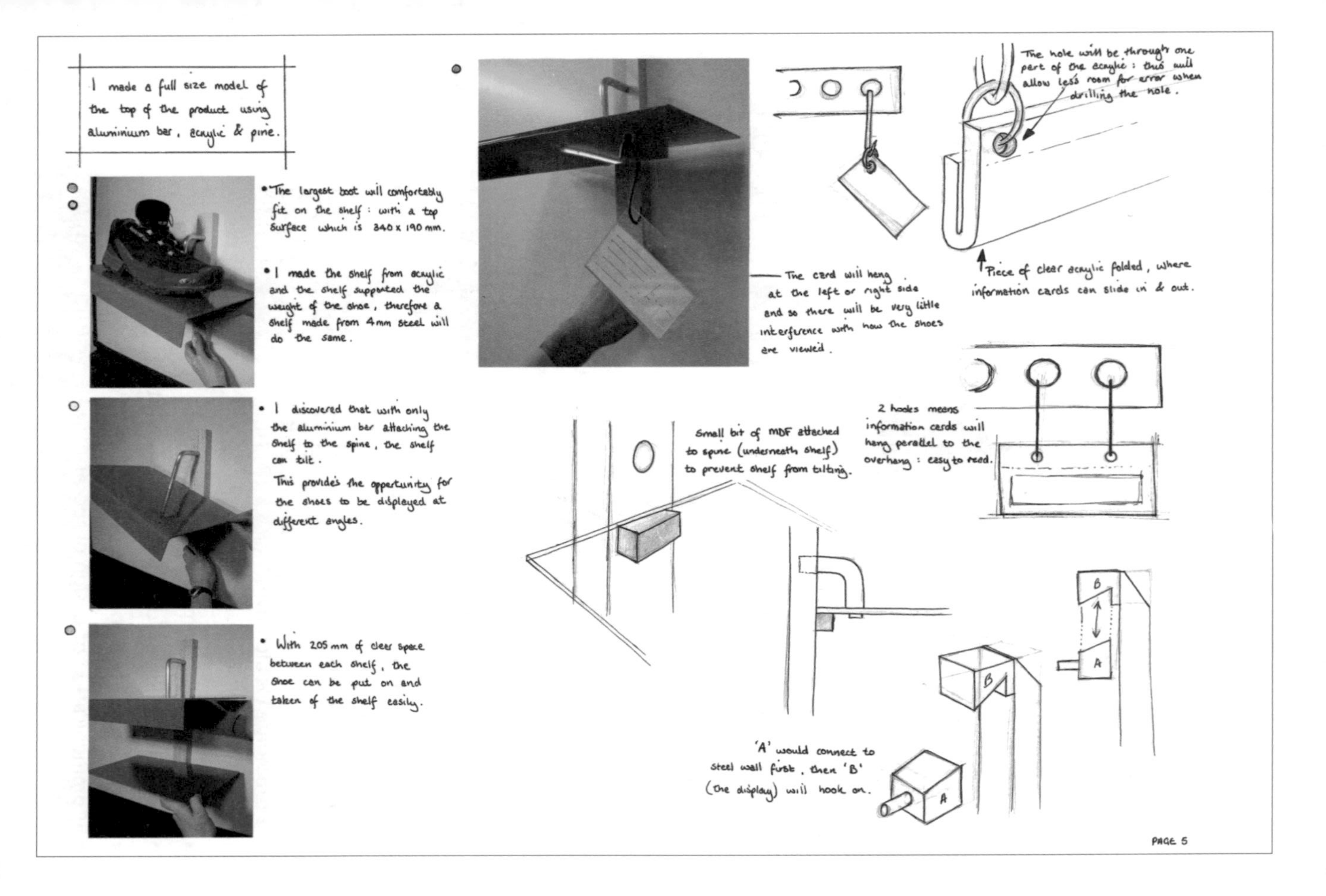
I made a full size model of the top of the product using aluminium bar, acrylic & pine.
• The largest boot will comfortably fit on the shelf : with a top surface which is 340 x 190 mm.
• I made the shelf from acrylic and the shelf supported the weight of the shoe, therefore a shelf made from 4mm steel will do the same.
• I discovered that with only the aluminium bar attaching the shelf to the spine, the shelf can tilt.
This provides the opportunity for the shoes to be displayed at different angles.
• With 205 mm of clear space between each shelf, the shoe can be put on and taken of the shelf easily.
The card will hang at the left or right side and so there will be very little interference with how the shoes are viewed.
The hole will be through one part of the acrylic : this will allow less room for error when drilling the hole.
Piece of clear acrylic folded, where information cards can slide in & out.
2 hooks means information cards will hang parallel to the overhang : easy to read.
Small bit of MDF attached to spine (underneath shelf) to prevent shelf from tilting.
B
A
B
A
'A' would connect to steel wall first, then 'B' (the display) will hook on.
PAGE 5

Tip

Don't leave model making to last. Models offer more potential when used as a development tool rather than for mere presentation.

Here the candidate highlighted unforeseen problems and created potential for meaningful development.

Modelling highlighted problems relating to the display of information, but also offered the potential to solve this problem.

Modelling gave rise to more detail relating to the way the information was held and positioned on the shelf.

Modelling has also provided the candidate with a better insight into some of the construction problems and fixing it to the wall.

This shows that the candidate is refining their initial idea towards a final solution.

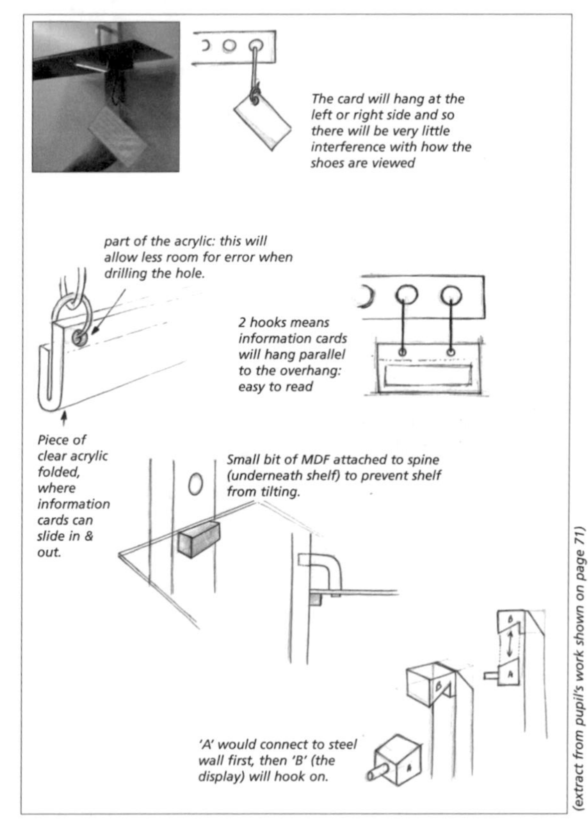

(extract from pupil's work shown on page 71)

The candidate would have gained credit for:

- Development towards a design proposal
- Communication of ideas and development
- Recording and justification of decisions taken
- Communication of design proposal.

Tip

Making models is an effective means of developing and presenting ideas. Models offer the potential to view ideas in 3D, generating valuable information on a wide range of issues. Choose the correct material and model type that suits the information you require.

• The largest boot will comfortably fit on the shelf: with a top surface which is 340 x 190 mm.

• I made the shelf from acrylic and the shelf supported the weight of the shoe, therefore a shelf made from 4 mm steel will do the same.

• With 205 mm of clear space between each shelf, the shoe can be put on and taken off the shelf easily.

• I discovered that with only the aluminium bar attaching the shelf to the spine, the shelf can tilt.

This provides the opportunity for the shoes to be displayed at different angles.

(extract from pupil's work shown on page 71)

Here the candidate opted to make a prototype as their development was centred around the construction and function of the design at this point.

The model offered the potential to test out theories on size and strength.

Modelling highlights potential that is not always obvious from sketching alone.

Experimentation is more likely through modelling. What may have been an accident can often lead to an improved idea.

This shows that the candidate is exploring a wide range of solutions and provides evidence on relevance, manufacture and materials.

The candidate would have gained credit for:

- Development towards a design proposal
- Communication of ideas and development
- Recording and justification of decisions taken
- Communication of design proposal.

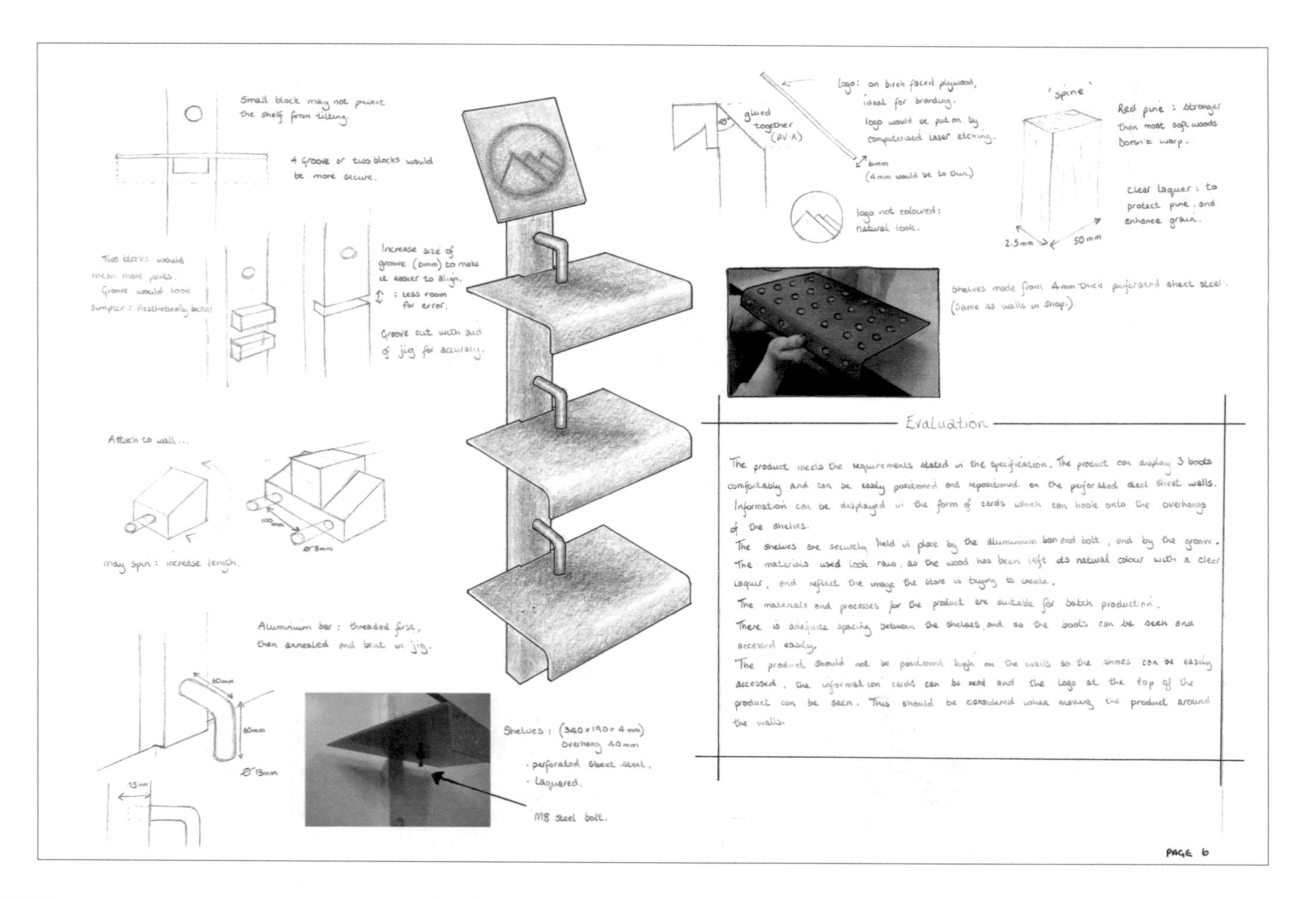
Small block may not prevent the shelf from tilting.
A Groove or two blocks would be more secure.
Two blocks would mean more parts. Groove would look simpler : Aesthetically better
Increase size of groove (6mm) to make it easier to align.
: Less room for error.
Groove cut with aid of jig for accuracy.
Attach to wall...
may spin : increase length.
100mm
Ø 8mm
Aluminium bar : threaded first, then annealed and bent in jig.
60mm
60mm
Ø 13mm
15mm
Shelves : (340 x 190 x 4 mm) Overhang 40mm
- perforated sheet steel.
- Laquered.
M8 steel bolt.
glued together (PVA)
45°
Logo: on birch faced plywood, ideal for branding.
logo would be put on by computerised laser etching.
6mm (4mm would be to thin.)
logo not coloured: natural look.
'spine'
Red pine : stronger than most soft woods Doesn't warp.
Clear laquer : to protect pine, and enhance grain.
2.5mm
50mm
Shelves made from 4mm thick perforated sheet steel. (Same as walls in shop.)
Evaluation
The product meets the requirements stated in the specification. The product can display 3 boots comfortably and can be easily positioned and repositioned on the perforated steel sheet walls.
Information can be displayed in the form of cards which can hook onto the overhangs of the shelves.
The shelves are securely held in place by the aluminium bar and bolt, and by the groove.
The materials used look raw, as the wood has been left its natural colour with a clear laquer, and reflect the image the store is trying to create.
The materials and processes for the product are suitable for batch production.
There is adequate spacing between the shelves, and so the boots can be seen and accessed easily.
The product should not be positioned high on the walls so the shoes can be easily accessed, the information cards can be read and the logo at the top of the product can be seen. This should be considered when moving the product around the walls.
PAGE 6

> **Tip**
>
> Make sure you leave enough time to present your finished proposal.
>
> This could include 2D or 3D rendered drawings, models or a combination of all three.
>
> Remember, time is important – don't spend too long producing finished models.

Here the candidate has produced a fully rendered drawing to describe the form, texture and materials. Additional drawings have been used to detail separate parts of the design and reference them to the specification. The existing model was also adapted to create a better understanding of the proposed solution.

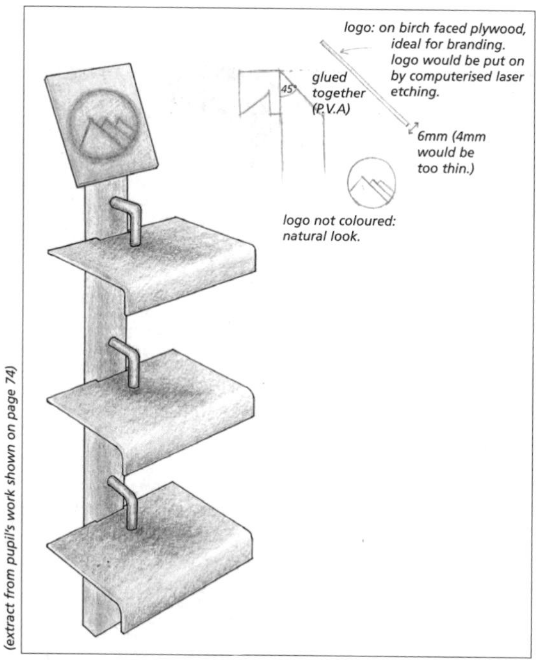

(extract from pupil's work shown on page 74)

The final solution has clear visual links with the initial idea.

A modern aesthetic has been created through form and material to reflect the correct image.

Each part of the design has been fully developed and justified throughout the assignment.

Few questions have been left unanswered relating to the proposal's function, aesthetic or construction.

Each part of the design has been fully developed with clear paths, linking initial ideas through development to the final solution.

Candidates would have gained marks for:

- Communication of ideas and development
- Recording and justification of decisions taken
- Communication of design proposal.

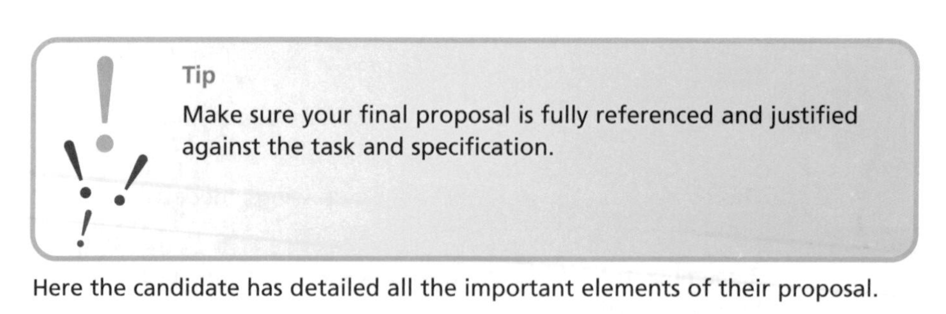

Tip

Make sure your final proposal is fully referenced and justified against the task and specification.

Here the candidate has detailed all the important elements of their proposal.

Separate sketches draw attention to details.

Some essential dimensions have been included.

Final decisions regarding methods of fixing and construction have been highlighted.

Tip: At times written information is the best form of communication. An honest evaluation should be included in order to ensure credit is awarded for relevance and to highlight a thorough understanding.

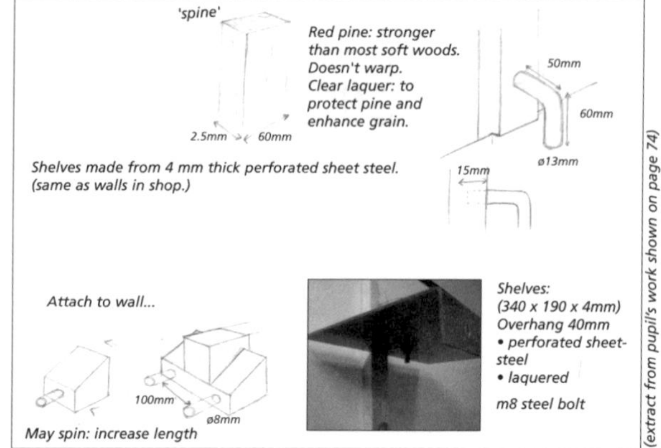

(extract from pupil's work shown on page 74)

The evaluation clearly explains the thinking behind the features, functions, materials, construction, manufacturing and environment which have helped shape and develop this proposal.

Evaluation

The product meets the requirements stated in the specification. The product can display 3 boots comfortably and can be easily positioned and repositioned on the perforated steel sheet walls.

Information can be displayed in the form of cards which can hook onto the overhangs of the shelves.

The shelves are securely held in place by the aluminium bar and bolt, and by the groove.

The materials used look raw, as the wood has been left its natural colour with a clear laquer, and reflect the image the store is trying to create.

The materials and processes for the product are suitable for batch production.

There is adequate spacing between the shelves, and so the boots can be seen and accessed easily.

The product should not be positioned high on the walls so the shoes can be easily accessed, the information cards can be read and the logo at the top of the product can be seen. This should be considered when moving the product around the walls.

(extract from pupil's work shown on page 74)

The candidate would have gained credit for:

- Communication of ideas and development
- Recording and justification of decisions taken
- Communication of design proposal.

4 The Written Paper

- ***Preparing for the written exam***
- ***Dividing the whole course into nine topics***
  1 *The design process*
  2 *Researching information*
  3 *Idea generation techniques*
  4 *Designing for people*
  5 *Communicating ideas*
  6 *Factors that influence design decisions*
  7 *Manufacturing systems*
  8 *Manufacturing processes*
  9 *Materials*
- ***Revising for question 1***

PREPARING FOR THE WRITTEN EXAM

At the end of your course, usually around May or June, you will be required to sit a two-hour examination which will test your knowledge of product design. This exam is worth 70 marks and will account for 50% of the total awarded for this course. Therefore it is important that you prepare thoroughly and have a revision plan which covers all the topics included in the course.

One of the main dangers is that you do not leave enough time to revise thoroughly, especially if you are sitting examinations in other subjects around the same time. Therefore writing out a timetable similar to that shown on page 11 will allow you to plan your study in advance. You will see from this timetable that the course has been broken down into nine topics. This makes revision more manageable and will give you the opportunity to revise whole sections of work in short, self-contained sessions.

The nine topics are:

1. The design process
2. Researching information
3. Idea generation techniques
4. Designing for people
5. Communicating ideas
6. Factors that influence design decisions
7. Manufacturing systems
8. Manufacturing processes
9. Materials

The content of each of the above topics is set out in the following pages and you will see that there is some duplication between topics. Don't let this worry or confuse you. It should allow you to put things into a broader context and help you to consolidate your knowledge.

You will notice that alongside each topic there is a page by page reading list that you can refer to from *Higher Product Design Course Notes* also published by Leckie & Leckie. This should allow you to revise as you go.

Prior to the final exam you may be required to sit a prelim which may or may not cover all the topics included in the course. Ask your teacher before the prelim which topics you should revise. Also, as you finish each week of your course, you can 'tick off' topics as you encounter them, allowing you to build up a picture of what you have done in class.

It is never too early to begin your revision for the written exam!

DIVIDING THE WHOLE COURSE INTO NINE TOPICS

Trying to revise a whole course can often seem a daunting task. Breaking the course down into topics, each with its own contents list, should be more manageable and should enable you to identify areas that you feel you are weak in and need more attention.

Throughout your course you will have been asked to carry out research, design products and produce design folios. It is unlikely that you will have studied each of the nine topics listed in such a structured and prescribed way. However, as you read through each topic list you will already be familiar with many of the headings and be able to put them into the context of your own design work.

The page numbers beside each topic refer to the textbook *Higher Product Design Course Notes*.

The design process

(Pages 8, 10, 12, 14, 18, 20, 22, 40, 44, 50, 70, 72, 74, 80)

- ❑ Identifying needs, wants and problems
- ❑ Open and closed design briefs
- ❑ The design specification (technical, performance, market)
- ❑ Members of a typical design team
- ❑ Design development considerations
- ❑ Product planning and product strategy
- ❑ Product life
- ❑ Evaluation (user trips, observation, testing, comparison, check specification)
- ❑ A typical model for the design process
- ❑ Analysis
- ❑ Synthesis

Researching information

(Pages 54, 56, 58, 64, 66, 76, 78, 86, 102)

- ❑ Target market
- ❑ Market research
- ❑ Researching information
- ❑ Marketing or selling

Idea generation techniques

(Pages 24, 26, 28, 30, 32, 34, 36, 38, 42, 44)

- ❑ Morphological analysis
- ❑ Transfer
- ❑ Analogy
- ❑ Lateral thinking
- ❑ Thought showers
- ❑ Mind maps
- ❑ Lifestyle boards
- ❑ Mood boards

Designing for people

(Pages 54, 56, 58, 60, 62, 64, 68, 72, 80, 82, 86, 90, 110, 114)

- ❑ Aesthetics (shape, form, line, colour, proportion, contrast, texture, pattern)
- ❑ Ergonomics (anthropometrics, psychology, physiology)
- ❑ Defining the market
- ❑ Benefits of product design
- ❑ Fashion, style and fads
- ❑ Market trends
- ❑ Needs and wants
- ❑ Safety

Communicating ideas

(Pages 40, 48, 96, 196, 202, 204)

- ❑ Models
- ❑ Prototypes
- ❑ Systems diagrams
- ❑ CAD
- ❑ Types of drawings and their uses
- ❑ Rapid prototyping

Factors that influence design decisions

(Pages 12, 16, 18, 39, 46, 64, 72, 74, 76, 78, 80, 82, 84, 86, 88, 94, 98, 100, 102, 104, 106, 108, 110, 114, 120, 124, 126, 184, 186)

- ❑ The design brief
- ❑ Primary and secondary functions
- ❑ Specifications
- ❑ Market research
- ❑ Fashion and style
- ❑ Fitness for purpose
- ❑ Planned obsolescence
- ❑ Redundancy
- ❑ Technological opportunity
- ❑ Miniaturisation
- ❑ Ergonomics
- ❑ Environmental considerations (sustainable design, recycling)
- ❑ Consumer demands (technology push/consumer pull)
- ❑ Social expectations, social responsibilities and social behaviour
- ❑ Market opportunity, niche marketing
- ❑ Product life
- ❑ Research and development
- ❑ Product testing
- ❑ Safety and legislation
- ❑ Designing for manufacture
- ❑ Choosing a material
- ❑ The cost of production
- ❑ Economic considerations, costs (fixed and variable)
- ❑ Intellectual property

Manufacturing systems

(Pages 124, 186, 188, 190, 192, 194, 196, 198, 200, 202)

- ❑ The cost of production

- ❑ One off production
- ❑ Batch production (jigs, templates, patterns)
- ❑ Mass production
- ❑ Line production
- ❑ Cell production (flexible manufacturing)
- ❑ Rapid prototyping
- ❑ Just In Time production
- ❑ Sequential engineering
- ❑ Concurrent engineering
- ❑ CAD/CAM and CNC machining
- ❑ Automation
- ❑ Mechanisation

Manufacturing processes

(Pages 124, 140, 142, 144, 146, 148, 150, 152, 154, 162, 164, 166, 168, 170, 174, 182)

- ❑ Injection moulding (plastic)
- ❑ Extrusion (metal and plastic)
- ❑ Rotational moulding (plastic)
- ❑ Vacuum forming (plastic)
- ❑ Blow moulding (plastic)
- ❑ Laminating (plastic)
- ❑ Compression moulding (plastic)
- ❑ Metal turning
- ❑ Milling metal
- ❑ Die-casting (metal)
- ❑ Press forming (metal)
- ❑ Piercing and blanking (metal)
- ❑ Joining metal (welding, riveting, bolts, screws, adhesives)
- ❑ Sand casting (metal)
- ❑ Forging and drop forging (metal)
- ❑ Finishing metal (electroplating, paint, laquer)
- ❑ Wood turning
- ❑ Routing (wood)
- ❑ Spindle moulding (wood)
- ❑ Laminating (wood)
- ❑ Jointing (wood) – carcase and frame construction

Materials

(Pages 126, 128, 130, 132, 137, 138, 156, 158, 160, 172, 174, 176, 178, 180)

- ❑ Polythene (high and low density)
- ❑ Polyvinyl chloride
- ❑ Polystyrene
- ❑ Nylon
- ❑ Acrylic
- ❑ ABS
- ❑ Polypropylene
- ❑ Melamine formaldehyde
- ❑ Urea formaldehyde
- ❑ GRP (glass reinforced plastic)
- ❑ Carbon-fibre plastics
- ❑ Elastomers
- ❑ Mild steel
- ❑ Stainless steel
- ❑ High carbon steel
- ❑ Cast iron
- ❑ Brass
- ❑ Bronze
- ❑ Aluminium
- ❑ Duralumin
- ❑ Copper
- ❑ Tin
- ❑ Lead
- ❑ Zinc
- ❑ Beech
- ❑ Oak
- ❑ Ash
- ❑ Mahogany
- ❑ Teak
- ❑ Balsa wood
- ❑ Walnut
- ❑ Scots pine
- ❑ Red cedar
- ❑ Parana pine
- ❑ Spruce
- ❑ Plywood
- ❑ Medium density fibreboard (MDF)
- ❑ Blockboard
- ❑ Chipboard
- ❑ Hardboard
- ❑ Veneer

1 THE DESIGN PROCESS

EXAM EXAMPLE 1

1 Consider yourself in the role of a consumer watchdog. Give an outline description of four criteria you would use to evaluate a new domestic iron. (4)

This is a poor pupil answer.

I would evaluate ergonomics, comfort, safety and how much the iron costs.

Why is this a poor pupil answer?

The pupil has not given an outline description of any of the points and therefore has not answered the question. Also, comfort is an area that would be evaluated under the heading of ergonomics. This pupil has demonstrated only a limited knowledge of evaluation criteria. (1 mark)

This is a top pupil answer.

(i) Safety is an important criterion. Things to be considered are: How stable is the iron when sitting on the ironing board? How easily might the cord wear or fray?

(ii) Ease of use is important. Is it comfortable to hold? Can it be filled with water easily? Are the controls easy to understand?

(iii) What functions and features the iron has. Range of temperature control settings, spray and steam options can be evaluated.

(iv) Is the iron environmentally friendly? Two areas that should be considered are: How much power does it use? Can the materials be recycled?

Why is this a top pupil answer?

This pupil has selected four different criteria and described what can be evaluated in each. Clearly they have a deeper understanding of how to evaluate an iron. The answer is also set out in four separate parts. (4 marks)

EXAM EXAMPLE 2

2 Three aspects of the design process are listed below (a–c). What is the function of each and what should be specified in each? *(Higher Craft and Design, Question 2, 1996)*

(a) The initial need or client's specification (2)

(b) The product design specification (2)

(c) The product manufacturer's specification (2)

This is an average pupil answer.

(a) This tells everyone what the product should do. Things like how it should perform and how much it will cost.

(b) What the product must do, what the product must be and what the product must have should be explained.

(c) This should tell everyone how well it meets the health and safety requirements.

Why is this an average answer?

Each part of this answer (a–c) does not clearly explain what type of information should be given for each type of specification but it does demonstrate that the pupil knows there is a difference between each. (1 mark for each part)

This is a top pupil answer.

(a) At the outset of any design project it is essential to talk to the client and establish an understanding of what the product must do, what it should include and what market it is for. Both the designer and the client must be clear on things such as what benefits it should offer, what market segment it is having to compete in, who the consumers/users will be and what their needs are. They should also be sure that the product they propose is the most appropriate to fulfil these requirements. The client will want to specify certain things and will have certain constraints and restrictions. For example time will be a big factor and the designer needs to know what time limits are to be imposed. Costs are also a major consideration and the designer again must be aware of what constraints are involved when considering things like materials, construction methods, standard components, etc.

(b) The product design specification must deal with all that the product must do and how it will perform. For example a Technical

Specification will have to be written and should state exactly what the product will use in the way of standard components giving values, operating conditions and details of replacement and repair. It should also contain a marketing specification, which would give a direction on things like aesthetics, image, look and feel of the product as well as indicating who the anticipated user groups are. Also information on how the product is expected to be used and might be abused should be included. Information on the product's performance is expected. This might be noise levels, or the brightness of a light. Also what inputs are needed to make the product function properly.

(c) *The product manufacturer must give information on safety regulations and make sure that the product conforms to them. They also must give specific values and ratings of standard components such as batteries, bulbs, motors etc., detailed information on all materials used and processes used during the manufacture are essential.*

Why is this a top pupil answer?

The information for each part is specific and sufficiently detailed to show that this pupil understands the purpose of each type of specification and that there are differences between them. (Full marks)

EXAM EXAMPLE 3

3 A product designer will usually enlist the help of other experts during the design stages of a new product. With regard to the design of an electric toaster choose four experts that a designer may work with and briefly describe the help, advice and information they could provide. (4)

This is a good pupil answer.

(i) *The designer will have to ensure that the toaster is electrically safe and would benefit from discussing their ideas with an electrician.*

(ii) *The designer will also want to be sure that their idea will sell well in the market place and therefore will want to discuss all potential ideas with someone from marketing to be sure that the ideas will compete in the market place with existing toasters from other companies.*

(iii) *The designer will want to make sure that the proposed solution can be manufactured and will discuss with a production engineer the design of each part and how easy they will be to make. Also the cost of production is important.*

(iv) With a product like an electrical toaster it is likely that many of the parts will be bought from other companies. Therefore it would be important to talk to suppliers and subcontractors.

Why is this a good pupil answer?

The pupil has set out the answer well in four distinct parts and chosen a different expert to talk about each time. For each one a good explanation of how they could help has been given. (Full marks)

This is a top pupil answer

(i) It is unlikely that the designer will be an expert in electrics. It would be useful for them to consult with an expert in this field who would advise them on what the health and safety requirements are to make a domestic product electrically safe for use in the home.

(ii) A product like an electric toaster will have lots of competition and it will be difficult to design such a product that will break into this highly competitive market. There is no point designing this product if no one will buy it. Therefore it would be useful for the designer to employ a market research team who could show sketches, drawings and models of the idea to potential customers and conduct a survey to find out if there would be enough interest in the toaster to justify going ahead with full production.

(iii) It is important the cost of production is kept to a minimum, therefore the design of each part should be discussed with a production engineer. It may be that with changes to the part the cost of making the mould and producing the part can be reduced. Also a decision may be taken to sub contract the manufacture of some parts to save money on buying machines and equipment.

(iv) Heating elements, plugs, cables, screws etc are all standard parts and would be bought from other suppliers. A production manager would be given the task of sourcing these parts and managing their purchase and delivery. This may involve Just In Time production methods.

Why is this a top pupil answer?

This pupil has written about the same issues as the previous pupil but has shown a much deeper understanding of each. For example, they have referred to health and safety issues, conducting a survey, mould making and JIT. If the question were worth eight marks this would still attract full marks.

2 RESEARCHING INFORMATION

EXAM EXAMPLE 1

1 In the past companies have manufactured and tried to sell their products – a selling policy. They are now adopting a marketing policy which is proving to be more effective. Discuss:
(a) the difference between selling and marketing (3)
(b) the implications of this change for both the consumer and the company. (3)
(Higher Craft and Design, Question 4, 1992)

This is an average pupil answer.

(a) Marketing means advertising a product and trying to compete with other similar products. Selling is when you put a product up for sale in a shop and just wait for someone to come and buy it. There is no advertising or marketing just a product for sale and customers can take it or leave it.

(b) The consumer is aware of greater choice if companies adopt a marketing policy. If they only sell a product the consumer may not be aware that the product exists and therefore buy something else.

Why is this an average pupil answer?

(a) Marketing does not mean advertising. This is only part of a wider strategy. The pupil does explain selling much better and shows that they know there is a difference. (1 or 2 marks could be awarded here)

(b) The implications for the consumer could have been explained further. The pupil has not discussed the implications for the company at all. (1 mark)

This is a top pupil answer.

(a) There are many differences between selling and marketing. Marketing can be described as a whole approach to product development, distribution and sales. It will take account of the correct marketing mix (Product, Price, Place and Promotion). Market research should establish consumer needs and wants and determine if the proposed product will be successful. Identifying the correct 'target market' is essential as this will influence the product's aesthetics and price amongst other things. Marketing also involves identifying appropriate

points of sale for the product to attract the 'target market group' and encourage them to buy the product at the price set.

Selling is much more straightforward and could be described as offering a product or service for sale with little or no regard for any of the points made above.

(b) The implications to the consumer now that companies are adopting a marketing policy rather than a selling policy are:
- better products which meet more individuals' needs
- products sold at easily accessible retail outlets
- products priced competitively

The implications of this change in policy for the company would be
- more time spent developing and designing product ideas to meet identified consumer needs
- more money spent on researching consumers' needs and wants
- being more aware of competition in the market place and responding much quicker to change.

Why is this a top pupil answer?

This is a thorough answer. Take note of how the answer to part (b) has been set out to answer the implications for the consumer and the company. Compare this to the average pupil answer to part (b). (Full marks awarded for both parts)

EXAM EXAMPLE 2

2 Describe two types of activity that are carried out during:
(a) desk research **(2)**
(b) field research **(2)**

This is a poor pupil answer.

(a) Looking at the computer and reading books.
(b) Doing surveys and testing things.

Why is this a poor pupil answer?

This is a very weak answer showing no understanding of any of the activities stated by the pupil and how they relate to research work for product design. The question asks for a description – no description has been given. (0 marks)

This is a top pupil answer.

(a) *Desk research means that the designer can work from resources and information close to their office or home. One example would be downloading relevant information about their design work from the web or finding something out such as information on materials from a web site. Another example would be reading articles from books or magazines that they own and have easy access to in the office or at home.*

(b) *Field research requires the designer to go beyond their home or office and seek out the information they require even if at first they don't know where to find it. A good example is asking questions and discussing with consumers their views on some area of design work that the designer wants information on. Another example would be going to a library to find specific information that they do not have access to in their office.*

Why is this a top pupil answer?

This pupil starts off their answers by describing what field research and desk research is. This makes it easy for them to put into context the examples of activities used for each. The pupil reminds the marker that they have given two examples for each by saying 'Another example would be …'. (Full marks awarded)

3 IDEA GENERATION TECHNIQUES

EXAM EXAMPLE 1

1 A professional designer will often use a variety of techniques to encourage divergent thinking in the search for innovative solutions to design problems.
Describe each of the following techniques
(i) Thought showers (2)
(ii) Morphological analysis (2)
(Higher Craft and Design, Question 7, 1995)

This is an average pupil answer.

(i) This means that lots of ideas are considered at the same time and nothing is rejected. Lots of lists are written down and talked about later.

(ii) Again lots of lists are written and words from each list are put together to help the person think of new ideas.

Why is this an average pupil answer?

It is clear that this pupil knows what the two techniques are and has probably used them at some point in their course. However, the explanations could be much fuller and the conditions under which the techniques are used should have been described. (1 or 2 marks could be awarded for part (i) and only 1 mark for part (ii))

This is a top pupil answer.

(i) Thought showers is a technique used to generate new ideas. It is best done in a group. This allows everyone to listen to each others ideas which should in turn help them to think of their own new ideas. No idea should be rejected and every one recorded onto paper. Once the exercise has been completed each idea in turn should be discussed and a decision taken by the group whether it should be accepted or rejected.

(ii) This is a much more structured and controlled technique used to create new ideas. This can be done alone. Headings are written down on paper and words listed below which relate to each heading.

Random words are selected from each heading and combined to see if they help create a new idea or thought.

Why is this a top pupil answer?

This is a much more structured and thoughtful answer which describes each process more clearly. The pupil has also described the conditions under which each technique is best used. (Full marks)

EXAM EXAMPLE 2

2 Explain what is meant by the term 'Technology Transfer'. (2)

This is a poor pupil answer.

Technology transfer is where one idea is used in another product.

Why is this a poor pupil answer?

This question is worth 2 marks and asks for an explanation. It is unlikely that this pupil will ever gain 2 marks in a Higher exam for writing just eleven words. It may be difficult to explain what a term like 'Technology Transfer' means so take the time to give an example to illustrate your explanation. (At best 1 mark might be awarded)

This is a top pupil answer.

Technology transfer is when an idea or part of an idea from one product is used to help the design of another product. An example of this is the laser technology used in CD players. This technology was not developed for home entertainment. It was developed for military or medical use and its potential for use in CDs was then recognised. The technology was adapted and then included in CD players at a much reduced cost.

Why is this a top pupil answer?

The explanation was much better and it was backed up by a good example to show that the pupil understands the principle of technology transfer. (Full marks)

EXAM EXAMPLE 3

3 Picture boards are used to help generate and develop ideas. Describe how a designer would use:
(a) a mood board (2)
(b) a lifestyle board (2)

This is a poor pupil answer.

(a) A mood board is a picture board that shows the mood of the designer. He will use it to help his design work.

(b) A lifestyle board shows pictures of people's lifestyle and these pictures will also help a designer to come up with ideas.

Why is this a poor pupil answer?

(a) It is clear that this pupil does not have a full understanding of what a mood board is. It should not reflect the mood of the designer but it will help inspire the designer during the process. (At best 1 mark might be awarded)

(b) The lifestyle board should reflect the lifestyle of those in the intended market group. The pupil has said it reflects people's lifestyle which is a weaker response. (1 mark could be awarded here)

This is a top pupil answer.

(a) A mood board should show a collection of pictures that reflect a particular mood. For example happy, fun, exciting, full of life. This would provide the designer with images, colours, shapes and textures that could be used in the design of a product that was to appeal to a happy, fun type of market group. Maybe someone who enjoyed extreme sports.

(b) A lifestyle board will show pictures of someone's lifestyle. This might be the clothes they would normally wear, the house they live in and the car they drive. Any picture about their life could be included. A designer would use these images to help design a product that would appeal to that market group of people.

Why is this a top pupil answer?

Both answers are well explained and give appropriate examples of what would be included in each picture board. This helped the pupil describe how the designer would use this information particularly in answer (a) where the example of someone who likes extreme sports is used. (Full marks would be awarded for both answers).

4 DESIGNING FOR PEOPLE

EXAM EXAMPLE 1

1 Ergonomics can be regarded as the study of each of the following:
(a) Anthropometrics
(b) Physiology
(c) Psychology
Describe what is meant by each of these in relation to ergonomics. (6)

This is an average pupil answer.

(a) Anthropometrics is about measurements of the body. It is where a designer looks up a table or a chart to find the correct size they need.

(b) Physiology is about what the body can do and how it can perform.

(c) Psychology is about people's minds and what they think about the product they are looking at.

Why is this an average pupil answer?

This is a typical sort of answer that a pupil might give. It shows some knowledge of each of the three areas but does not really provide enough depth. None of the answers given describe any of the three areas in relation to ergonomics. This would have been easier to do if some explanation of ergonomics was given and an example was used to illustrate what is meant. (3 out of 6 would be a fair assessment of this answer)

This is a top pupil answer.

Ergonomics is the study of human beings and how they interact with their environment. It can be regarded as having three areas of study – anthropometrics, physiology and psychology.

(a) Anthropometrics is a collection of measurements and data about the human body which is presented in a reference book which can be accessed by anyone. It provides measurements for males and females between the 5th and 95th percentile for all ages. For example it would be possible to look up the popliteal height for a 5th percentile thirty-year-old female. This measurement can be used as a starting point for a designer who is designing a dining chair. It will not ensure that the chair is comfortable but will provide a starting point for design.

(b) *Physiology is an area of study which looks at how the body moves and its limits and restrictions. For example it would be useful when designing a hand held product, to know how the hand and fingers bend and move and what their limitations are in terms of reach, strength and flexibility. This is an area of ergonomics which is best investigated through modelling and interacting with products.*

(c) *The first impression of a product is very important if a consumer is going to buy it. Therefore its aesthetics, style and appearance will have an important influence on whether someone likes it or not. The feel good factor a product gives is psychological and this is an important part of ergonomics. As well as liking the product it should be easy to understand how to use it and interact with it. The product cannot speak to the consumer so it should give enough visual information to make the user understand how to use it. This is also an important psychological message that the product should give.*

Why is this a top pupil answer?

These three answers are much more in depth and do give a description. This pupil spent some time at the beginning writing down what ergonomics is. Although no marks would be awarded for this it makes answering the next three parts much easier. Examples of products are used in parts (a) and (b) which clearly demonstrate the pupil's understanding. This was not done for part (c) but for 2 marks the description is detailed enough to stand on its own. (Full marks awarded)

EXAM EXAMPLE 2

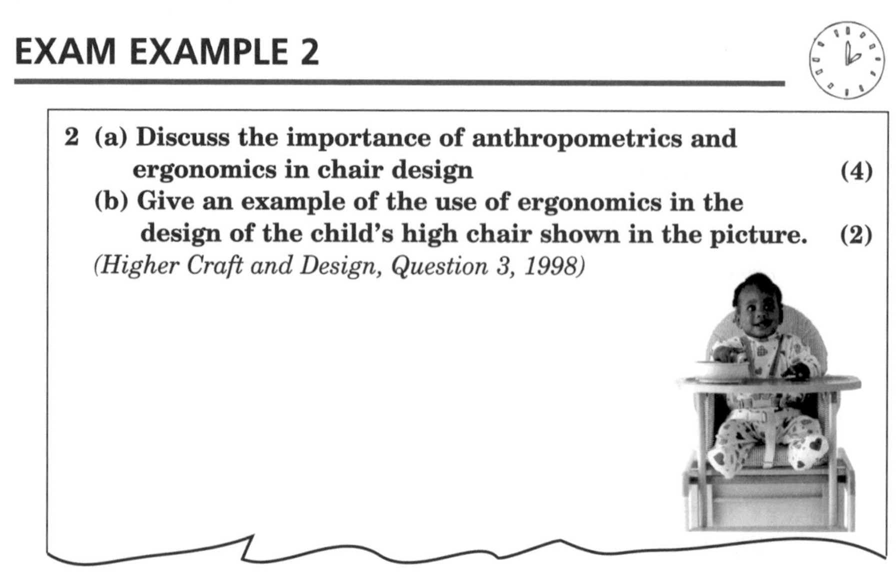

2 (a) Discuss the importance of anthropometrics and ergonomics in chair design **(4)**

(b) Give an example of the use of ergonomics in the design of the child's high chair shown in the picture. **(2)**

(Higher Craft and Design, Question 3, 1998)

This is a poor pupil answer.

(a) Anthropometrics is the collection of information and data relating to the human body. This information is presented in tables giving a range of measurements from the 5th to the 95th percentile of the population. It is important that this information is available to designers and that it is interpreted correctly and applied to the product being designed. For example in deciding the height of a shelf in a supermarket the 5th percentile of reach for women should be used. This would ensure that everyone within the 5th percentile of women and the 95th percentile of men could reach the shelf comfortably. There will be those outside of this range who will have great difficulty reaching the shelf and seeing what is on it but it is difficult to design one product that suits everyone. Ergonomics is the study of humans and how they interact with their environment.

(b) Ergonomics will have to be considered in the design of the children's high chair shown. Babies will wiggle and spill drinks, hit tables, rock, try to climb out of things and generally behave in a way that adults wouldn't. Therefore a designer must ensure that the chair is stable, strong and able to withstand a fair bit of abuse. It should be made from easily cleaned materials. The seat should be soft and comfortable and straps are included to secure the baby. The seat shown has a padded insert and straps. It looks stable and robust. The food tray is big enough and close enough to the baby to minimise drink spillage going over the baby.

Why is this a poor pupil answer?

(a) What is written is true but it does not relate to a chair, therefore cannot be given full marks. It does not describe the importance of either issue with regard to chair design. (0 marks for this answer)

(b) Again, good points are made but the answer does not clearly give examples of how ergonomics has been considered in the design of the baby's high chair. It refers mainly to how the baby would use the chair. (1 mark for this answer)

This is a top pupil answer

(a) If ergonomics is not considered carefully in the design of chairs it is likely that the chair will not provide the level of comfort required. Designers need to know what kind of chair they are designing and what the desired position of the body is to be when seated. Once this has been decided and the degree of comfort and support required from the chair has been established, the designer can access appropriate

anthropometric data. The 5th to 95th percentile range should be used for males and females and the appropriate size selected such as the popliteal height, angle of the back rest and hip width.

(b) The baby looks comfortable in the chair. Its arms rest comfortably on the table. Clearly the position and height of this has been considered to allow the baby to eat from the table and rest its arms on it. There is plenty leg room and enough for the baby to grow further. The distance between the table and the back of the chair has been considered to allow the baby easy access into the chair. The width of the seat also allows the baby to sit comfortably and there looks to be enough room to allow the baby to grow and still fit the chair.

Why is this a top pupil answer?

(a) The pupil has made relevant points in the context of a chair. They have shown that they understand the need to establish what type of chair it is and how the body is expected to interact with it. It is clear that they understand that only selected relevant sizes should be used from supplied anthropometric data. (Full marks for this answer)

(b) The pupil makes several points that show an understanding of some of the ergonomic issues the designer had to deal with during the design stages. (2 marks for this answer)

EXAM EXAMPLE 3

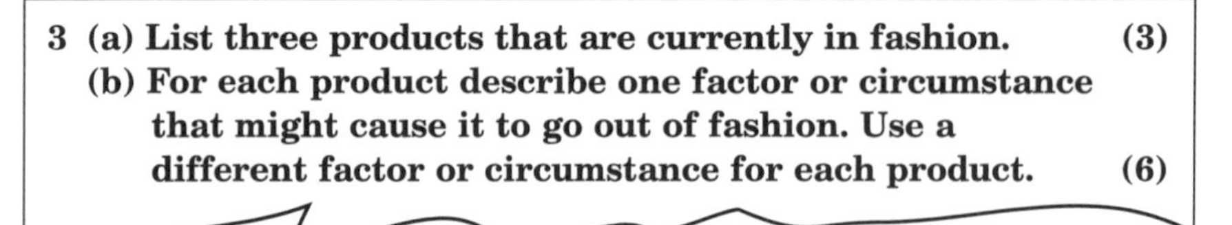

3 (a) List three products that are currently in fashion. (3)

(b) For each product describe one factor or circumstance that might cause it to go out of fashion. Use a different factor or circumstance for each product. (6)

This is a poor pupil answer.

(a) An ipod, a Nokia mobile phone and a .....

(b) An ipod might go out of fashion if another newer MP3 player was introduced that had more features on it. Mobile phones are changed and updated regularly when newer slimmer phones with better features are introduced.

Why is this a poor pupil answer?

(a) The pupil has only listed two products. It is important that you give three products if you are asked, particularly when the second part of the question

relies on you being able to write about the three you have chosen. As a last resort it is better to include any product than none at all even if you are not sure that the product is currently in fashion. (2 marks awarded)

(b) The reasons given for the two products going out of fashion are too similar and therefore marks can only be awarded for one. Also, because only two products were listed in part (a) the pupil cannot write anything in this part. If the answer given in part (a) was wrong it would still be possible to gain marks for it in part (b). (2 marks awarded for this answer)

This is a top pupil answer.

(a) (i) An MP3 player is a fashionable product today.

(ii) A mobile phone that takes digital photographs and can access the internet.

(iii) Training shoes are also fashion items.

(b) (i) MP3 players can go out of fashion if new ones are introduced that have more memory space and have more functions and features. Also their size, shape and colour can change which can date existing MP3 players and make newer ones more desirable.

(ii) Mobile phones are constantly changing. Thinner slimmer phones which are more sleek looking can make older bigger phones less appealing to users.

(iii) Training shoes have been a fashion item for a long time. New materials on the shoe and the sole can make them more attractive to teenagers. Slip on shoes with Velcro fasteners are now more fashionable than shoes with big tongues and heavy soles which were in fashion years ago.

Why is this a top pupil answer?

(a) Three examples have been given. The additional information about the mobile phone is useful.

(b) The pupil has given a thorough answer to part (i) and given more than one example. This has made it difficult to give different examples for the other two. It is likely that 1 mark would have been awarded for part (i) improved memory; 1 mark for part (ii) newer shape and aesthetics; and 1 mark for part (iii) improved materials and different methods of fastening. Also, as a suitable description was given about each a further 1 mark for each part is likely to have been awarded. (Full marks for this answer)

5 COMMUNICATING IDEAS

EXAM EXAMPLE 1

1 Designers often use three dimensional models to communicate information about a product to a variety of people.
(a) Describe three examples of models they might use. (3)
(b) For each example write down what type of information could be found and to whom it would be useful. (3)

This is an average pupil answer.

(a) Designers could use clay models, MDF models and card models.
(b) Clay models can be used in the early stages of design work to give information about shape. MDF models are often used to present a final idea and paper models are useful at the early stages before the design work starts.

Why is this an average pupil answer?

(a) The three examples used are correct but each would benefit from a fuller description. The question asks that three examples are described. (Possibly 2 marks could be awarded)

(b) The pupil has only explained what type of information could be found from a clay model and has not made any reference for any of the models to whom the information would be useful. (At most 1 mark could be awarded here)

This is a top pupil answer.

(a) (i) A wet clay model can be used. Detail can be easily marked on the clay with tools.
(ii) Styrofoam models can give more precise detail and can be made in parts then glued together.
(iii) MDF models are usually presentation models which can be sprayed or painted to show much more precise detail.

(b) (i) A designer would find a clay model most useful. They could quickly make decisions on shape, proportion and aesthetics as well as ergonomics.
(ii) Styrofoam models would be useful for the designer to talk to other designers with or other members of the design team. It would give more precise information on shape and size and discussions might take place on how the product could be manufactured.

(iii) An MDF model would be used as a presentation model to a client. It would be as close to the proposed solution as is possible and be sprayed to look like the real thing. It would show the client exactly what it would look like and any changes decided upon before going into full production.

Why is this a top pupil answer?

(a) The answer has been well set out into three parts which makes it easier to mark. Good examples have been given with some additional information used to describe the sort of model. The pupil has been careful not to just list three examples. (Full marks could be awarded)

(b) The pupil has answered both parts of the question for each of the three examples: what type of information could be found and who would find it useful. The answer has been divided into three shorter parts rather than just one long answer. (Full marks could be awarded)

EXAM EXAMPLE 2

2 (a) List six types of drawing used to illustrate three dimensional design.
(b) Which type of drawing would be considered most appropriate to communicate an idea to a client who had no formal graphic training? Justify your choice. (3)

(Higher Craft and Design, Question 10, 1991)

This is an average pupil answer.

(a) Perspective drawing, isometric drawing, 2 dimensional drawing, free hand sketching, exploded drawings, full size drawings.

(b) Perspective drawing would be the most appropriate for someone who had no formal graphic training because they could see the design as it really is.

Why is this an average pupil answer?

(a) Freehand sketching is not really specific enough in this list. Two dimensional drawing is also too general but may be acceptable. Full size orthographic drawings would have been a stronger answer than just full size drawings. (Probably 2 marks would be awarded for this list)

(b) Perspective drawing is the correct answer and would be given 1 mark. The reason given is again too general and a fuller explanation would be required for the remaining 2 marks. (2 marks could be awarded for this answer)

This is a top pupil answer.

(a) Six types of drawing that could be used are (i) orthographic (ii) isometric (iii) perspective (iv) exploded views (v) sectional views (vi) solid modelling on a computer.

(b) A colour perspective would be the most appropriate because that is the way we see things. Even isometric which is close to perspective has a distortion caused by parallel lines. Perspective drawing is a three dimensional drawing which will provide clear information about the product without needing to refer to different views.

Why is this a top pupil answer?

(a) Six clear examples are listed. They are all different and correctly named. (Full marks could be awarded)

(b) Perspective is the correct answer. The pupil justifies this decision by giving reasons for selection and reasons why others should not have been selected. (Full marks could be awarded)

EXAM EXAMPLE 3

3 (a) Describe the differences between a model and a prototype. (2)

(b) Describe how a prototype could be used to test safety. (2)

'Rapid prototyping' is increasingly used in the production of prototype models.

(c) Describe how rapid prototyping models would be produced and their use in the development of a product. (4)

(Higher Craft and Design, Question 4, 2002)

This is a poor pupil answer.

(a) A model is not something that is real it is only used for testing. A prototype can be used in real life situations and is expensive to make.

(b) A prototype can be used to test things like strength and stability and decisions can then be made on these things before the product goes into full production and lots of money is spent on it.

(c) Rapid prototypes are produced by lasers in a resin bath. They are much quicker and cheaper to produce than real models and are becoming more popular with designers.

Why is this a poor pupil answer?

(a) A model is real and can also be expensive to make. The difference between the two has not been fully explained. (1 mark)

(b) What the pupil has said is correct but it does not describe how a prototype could be tested for safety. (0 marks)

(c) 1 mark could be awarded for saying that rapid prototypes are produced by lasers in a resin bath but there has been no explanation of how they could have been used in the development of a product. (1 mark could be awarded)

This is a top pupil answer.

(a) A prototype is a working model of the real thing. A prototype will perform the same way as the finished product should. For example a car manufacturer may make one prototype car and test it under the same conditions as the real car may experience. It will be expected to perform in the same way. This would be done before the car was mass produced. A model does not need to perform the same way as the finished product. It is used to evaluate aesthetics mainly.

(b) Once a prototype has been made it needs to be tested under the same conditions as the finished product is to be used in. This often means making up a test rig or simulating these conditions in some way. An example may be a baby's high chair. A dummy or a doll weighing the same as the baby could be sat in the chair as it was knocked and rocked to test for stability.

(c) Rapid prototyping is a process where real models are produced directly from a computer drawing. There are several ways this can be done but the most common amongst product designers is stereolithography. This requires a resin to be enclosed in a container alongside a table which can move up and down. Lasers which are controlled by a computer are fired onto the table. The resin then becomes solid and an accurate model is produced. This three-dimensional model can be used as a one off to test for things like ergonomics. It is the actual size and can perform the same as the real model.

Why is this a top pupil answer?

(a) The difference between the two has been better explained. This has been backed up with an appropriate example. It is clear this candidate has a deeper understanding of the difference between the two. (2 marks)

(b) This candidate knows that the prototype is tested under the same conditions and that these conditions need to be recreated somehow. The example helps the candidate describe how a very specific aspect of safety could be tested. (2 marks)

(c) A fairly good description of stereolithography has been given although some detail has been missed. A reasonable description of how the rapid prototype could be used has also been given. This answer does demonstrate that the pupil has knowledge beyond a basic level. (Probably full marks would be given)

6 FACTORS THAT INFLUENCE DESIGN DECISIONS

EXAM EXAMPLE 1

1 (a) What do you understand by the concept designed obsolescence? (2)
(b) Give two examples of products which illustrate this concept. (2)
(c) What effect does designed obsolescence have on:
(i) the environment (2)
(ii) the consumer (2)
(iii) the national economy? (2)

(Higher Craft and Design, Question 8, 1993)

This is a poor pupil answer.

(a) Designed obsolescence is when a designer designs a product to break down or wear out so as to encourage consumers to buy a new one.

(b) Washing machines and motor cars.

(c) (i) It creates pollution and waste because there will be products that nobody wants and they will have to be disposed of.

(ii) The consumer might feel that they are being ripped off.

(iii) The national economy will get bigger because lots of money will have been spent.

Why is this a poor pupil answer?

(a) This may have been true in the past but companies cannot do this now because there is so much competition and they have their reputation to consider. It may be possible to award 1 mark here but the pupil has not shown a full understanding of what designed obsolescence is by choosing this option. (Maybe 1 mark could be awarded for this example)

(b) Due to the poor answer in part (a) neither of these examples would be correct in today's market place. (0 marks)

(c) (i) Perhaps more could have been said about this for 2 marks. What is written is correct. (Probably worth full marks)

(ii) This answer would have to be expanded. The pupil's terminology of how the consumer might feel is not good. (This answer would struggle to get any marks)

(iii) It is clear from this answer that the pupil has no real idea about the national economy. (0 marks would be awarded)

This is a top pupil answer.

(a) Designed obsolescence is a strategy used deliberately by designers to outdate products or prematurely end their useful life. An example of this is computers. Much of the technology built in to a computer and many of the facilities it offers become outdated very quickly. In some cases it is often true that a newer faster more sophisticated model is already waiting to go on sale before its predecessor hits the shops. Many consumers want more up to date models and will sell or exchange their existing one for a newer version. The longer they hold on to their existing design the more obsolete the technology will become and it will be more difficult to run new software packages.

(b) Product 1 Computers; Product 2 Mobile phones, Training shoes

(c) (i) As the demand for newer more up to date products increases so the pressure to dispose of old products in an environmentally friendly way increases. This can lead to pollution, waste and litter. Even the disposal of the old product can pollute the environment through its collection and recycling.

(ii) The consumer is often caught in a trap of keeping up with the Jones's. There can be an unnecessary pressure placed on the consumer to buy the new model believing that the old one is outdated and unfashionable.

(iii) Designed obsolescence can lead to a demand for newer more up to date products. This means that the market is unlikely to become saturated and that there will be a demand to manufacture products. This will boost jobs. The economy is boosted by low unemployment and by high sales. Industry can respond through sustained production. As demand for new products grows so does the need for consumers to borrow money. [SA]

Why is this a top pupil answer?

(a) Computers are used as a good example to help the pupil explain their answer. It is much easier with a question like this to use a product to help the explanation. (Full marks)

(b) Appropriate examples have been chosen. (Full marks)

(c) The answer is well laid out and divided into three parts. Appropriate examples have been used and it is clear that the pupil has an understanding of all three issues. (Full marks)

EXAM EXAMPLE 2

2 Give two examples of products that have been developed as a result of technological opportunity. (2)

This is an average pupil answer.

CD players have been developed as a result of technological opportunity. This is because the laser inside them was not developed and produced for CD players. It would have cost too much money. Lasers were developed for space research or for military purposes and other manufacturers saw the opportunity to use that technology to make CD players. Mobile telephones are another example.

Why is this an average pupil answer?

The first part of this answer is excellent. It is well explained and shows a reasonable understanding of how the opportunity to use existing technology to develop another product can happen. The mobile telephone does not give a clear example of technological opportunity and requires some explanation. (1 mark awarded)

This is a top pupil answer.

(i) *There are many examples of household products which have been developed as a result of technological opportunity. Teflon is a non stick material used to coat pots and pans. Originally this material was developed through space research but the opportunity and benefits of using it on domestic cook ware was recognised.*

(ii) *Integrated chips used on powerful computers are now being reclaimed when the computer is no longer able to be used. They are reused on products like toys which may only need to use a fraction of its capability. This is an opportunity to reuse existing technology instead of disposing of it.*

Why is this a top pupil answer?

This answer has been well set out into two parts. Good examples have been given each with enough information to show that the pupil has a good understanding of the concept of technological opportunity. (Full marks awarded)

EXAM EXAMPLE 3

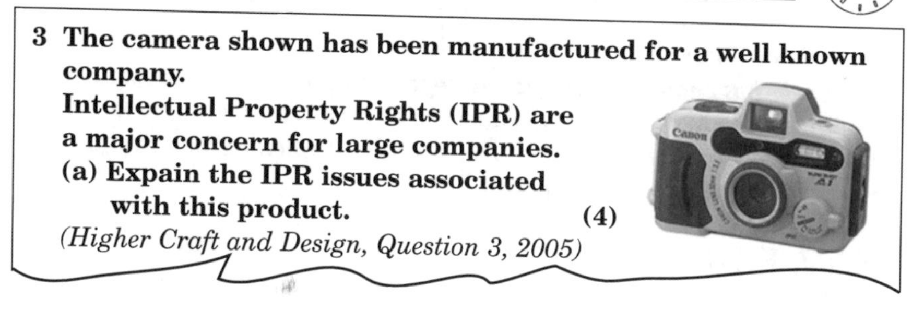

3 The camera shown has been manufactured for a well known company.
Intellectual Property Rights (IPR) are a major concern for large companies.
(a) Expain the IPR issues associated with this product. **(4)**
(Higher Craft and Design, Question 3, 2005)

This is a poor pupil answer.

This camera is someone's idea and they need to protect it in case someone steals it and says it is theirs. They can do this by taking a photograph of it and putting it in an envelope and posting it. Then if anyone says that the idea is theirs the designer has proof in a sealed envelope that it is not.

Why is this a poor pupil answer?

This question is worth 4 marks and therefore the marker will be looking for at least four points to be made in the answer. The one example given is not well described and is hardly appropriate for the camera shown. All the points the pupil makes should be appropriate for the product shown in the photograph provided. (1 mark might be given for a general understanding shown by the pupil)

This is a model pupil answer.

There are several IP issues that need to be considered by the company that designed this camera.

(i) The company has a trade mark which immediately identifies it as being different from other competitors. The company will have registered this trade mark and this prevents other companies from using it. It also means that customers can rely on the quality product that they associate with their products.

(ii) The design will have to be registered with the patents office to prove that it is theirs.

(iii) The designer who designed it may have signed a contract passing ownership of the idea to the company.

(iv) The company or the designer will have to keep safely all the original design drawings to prove that it is theirs.

(v) It would also be necessary for the company to check that they are not

infringing anyone else's idea before they manufactured the product. If they did this might mean that they were sued.

Why is this a top pupil answer?

This answer has been well set out into separate parts. Each point made is different and shows a good understanding of Intellectual Property Rights. The points made also refer to the product shown in the picture therefore the pupil has answered the question asked. (Full marks)

EXAM EXAMPLE 4

> **4 Use an example of a product with which you are familiar to describe what is meant by the term fitness for purpose. (3)**

This is an average pupil answer.

Fitness for purpose means that a product can do its job well. An example is a travel iron which is small and can fit in a suitcase.

Why is this an average pupil answer?

There is nothing wrong with this answer. The pupil has chosen an appropriate product but for 3 marks it may have been safer to write more about why the travel iron is fit for purpose. (2 or 3 marks could have been awarded here)

This is a top pupil answer.

A travel iron is a good example of a product that is fit for purpose. The travel iron is smaller than a normal household iron which makes it easier to pack in a suitcase. It is useful for ironing a shirt or t shirt on holiday but would be no good for doing the weekly ironing at home. The product was not designed for this purpose and is more suited to being used now and again for a light load.

Why is this a top pupil answer?

The same example of a travel iron has been used by this pupil. 1 mark would be given for choosing an appropriate example and 1 mark for each of the two points made. (Full marks)

7 MANUFACTURING SYSTEMS

EXAM EXAMPLE 1

1 Mechanisation can describe a procedure which is mechanically processed but is monitored by a human operator. Automation describes a similar procedure which is self monitoring.

(a) Describe one example of mechanisation (1)

(b) Describe one example of automation (1)

(c) State two implications that the move towards automation may have for each of the following:

(i) the manufacturing company (2)

(ii) the production workers (2)

(iii) the consumers (2)

(Higher Craft and Design, Question 7, 1993)

This is an average pupil answer.

(a) Mechanisation is when not everything is done automatically by machine and some things have to be done by hand. An example of this is using a CNC lathe.

(b) An assembly line in a factory where all the components are mounted on boards by machine.

(c) (i) It would mean that they will not need as many workers therefore they will save money.

(ii) The workers that are left will have to be retrained.

(iii) There will be no impact on consumers as they only buy the product.

Why is this an average pupil answer?

(a) There is enough information given to gain the 1 mark on offer. *(1 mark)*

(b) Not a full answer but enough for the 1 mark on offer. *(1 mark)*

(c) (i) Not enough information and not enough depth given to this answer. For 2 marks at least two points should be made. *(1 mark)*

(ii) The point made is valid but it is the only one made. *(1 mark)*

(iii) Not true. Automation will have an effect on the consumer. *(0 marks)*

This is a top pupil answer.

(a) One example of mechanisation is using a CNC lathe to produce small turned items. Most of the processes such as parallel turning, facing, tapering and screw cutting are done automatically and produced using co-ordinate information which has been put into the machine's computer system. However many other processes such as changing over the work piece blank and the cutting tool, setting the machine speed and feed rates still have to be done manually. Because of this manual input it cannot be regarded as being fully automated.

(b) Automation involves products or parts of products being produced entirely by computers and/or robots. Car manufacturing is becoming increasingly more automated. Parts are delivered within the factory by robots or buggies and positioned so as to enable robotic arms to pick them up and position them or do work on them. Car bodies and car shells are slowly moved along an automated line with parts being assembled, fixed and finished with no input from other workers.

(c) (i) Manufacturing companies that are moving towards automation will require to make a high investment in machinery, equipment and staff training. It may also require some reorganisation of internal work space and possibly more emphasis given to a clean environment.

(ii) The production workers may need to be retrained and in many cases may be laid off. This type of production involves fewer workers, some of whom will require to be highly skilled and highly trained. Others will require fewer skills and may be involved in store duties and other less technical jobs.

(iii) Consumers' expectations may increase. Certainly they will expect a quality product produced to the highest standards. In most cases they will expect the product to be available much quicker and at a lower cost. Products that are produced through automation often cost less and are often referred to as mass produced. Therefore the consumer cannot reasonably expect the product to be exclusive or have a designer label.

Why is this a top pupil answer?

All of these answers are good and comprehensive enough to gain full marks. In particular the information given in parts (a) and (b) is more than enough for the 1 mark on offer for each part.

EXAM EXAMPLE 2

2 Describe the conditions under which Just In Time Production (JIT) will not work efficiently. (3)

This is a poor pupil answer.

Just in time production will not work when the company is not organised well and no one is looking after the stock.

Why is this a poor pupil answer?

This answer is too general and shows no real knowledge of the demands of Just In Time production. The candidate is aware that this topic has something to do with stock control but does not demonstrate any real knowledge. It does not answer the question asked. (0 marks)

This is a top pupil answer.

Just In Time Production is a method of controlling stock and parts so that the company does not have to store too much and does not have to pay out money for things too early. It will not work if any of the five conditions cannot be achieved.

(i) Delivery of the parts to the factory on time is not guaranteed

(ii) The distance between supplier and the factory is too far away that there could be unforeseen delays

(iii) There is not a good stock control system in place

(iv) The parts being delivered are not good quality

(v) Production of the product could break down and machines could break.

Why is this a top pupil answer?

The pupil has set out the answer well and made answering it much easier by writing an introduction before stating the five points. It is clear that this pupil has revised well as all five conditions that would prevent JIT from being successfully implemented are given. It is worth writing more than required to ensure a better mark. (Full marks)

EXAM EXAMPLE 3

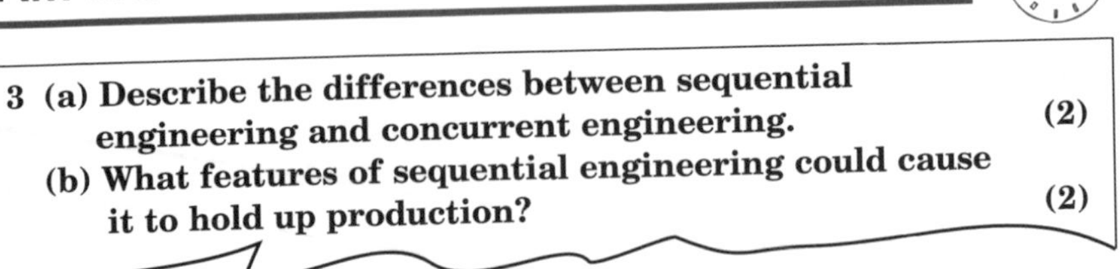

This is an average pupil answer.

(a) Sequential engineering is when things are manufactured in sequence one after another; concurrent engineering is when things are made at the same time.

(b) If there is a hold up with something then everything stops.

Why is this an average pupil answer?

(a) This answer does show that the pupil has some knowledge of the differences between the two methods and would be fine if the question had asked the pupil to state what is meant by the terms. The pupil is asked to describe the differences and for 2 marks it would have been better to illustrate the answer with an example. (1 mark would be awarded)

(b) Again this is a weak answer especially when 2 marks are on offer. The question asks for features, which implies at least two examples should be given. While what has been written is true, it has not been justified or explained. (1mark could be awarded but this is generous)

This is a top pupil answer.

(a) Sequential engineering is a method of manufacturing where each stage of the production is undertaken a bit at a time. It relies on one stage of the manufacture being completed before the next stage can begin. It can be compared to an assembly line production. Concurrent engineering is a method of manufacturing which is planned so as several stages of production are undertaken at the same time. This is a more modern approach which usually results in the product being manufactured quicker. An example might be a flat pack chair which has several parts. Each part could be made simultaneously by different people or by different companies before being packaged.

(b) The success of sequential engineering depends on each stage of the manufacturing being completed on time without any problems. If there is a hold up at any stage such as machines breaking, workers

being absent or a quality assurance issue, then production at the later stages would stop with those people involved having nothing to do.

Why is this a top pupil answer?

(a) This is a very thorough answer for 2 marks. The pupil has taken the time to describe the differences and a good example of the flat pack chair is given which helps the description. (Full marks awarded)

(b) A good answer which gives several reasons for sequential engineering holding up production. (Full marks awarded)

8 MANUFACTURING PROCESSES

EXAM EXAMPLE 1

1 Rotational moulding and injection moulding processes were used to manufacture the chairs shown below.

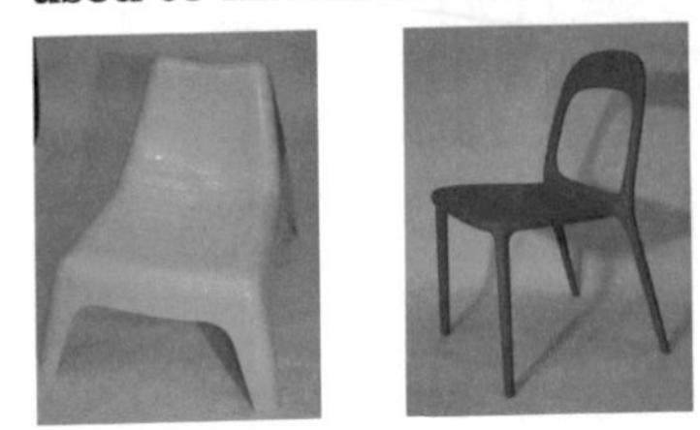

(a) What visual evidence would indicate which chair had been injection moulded and which had been rotational moulded?(2)

(b) Choose one of the chairs and explain why the process used is suitable for its manufacture. (2)

(c) State a suitable plastic for each process. (2)

This is an average pupil answer.

(a) The chair that had been injection moulded would have ejection marks on it and have tapered sides. The chair that had been rotational moulded would be hollow.

(b) Chair A was injection moulded because it is plastic and required a shiny finish.

(c) ABS could have been used for both chairs.

Why is this an average pupil answer?

(a) Both the points put forward for identifying injection moulding as a process are acceptable although they could have been more fully described. The question asks for visual evidence, therefore stating that the chair would be hollow for rotational moulding may not be accepted as a reason. (At least 1 mark would be awarded for this answer)

(b) The two reasons given are weak and could apply equally to other plastic moulding processes. (1 mark could be awarded)

(c) A different plastic would be required for each chair. (1 mark)

This is a top pupil answer.

(a) On the chair that had been injection moulded there would be evidence of ejection marks left by ejector pins when the chair had been released from the mould. There would also be an injection point mark where the sprue pin had been removed. Other features that should be looked for are flashing where the mould had separated and tapered sides. The

chair that had been rotational moulded would be hollow and have a thicker wall to it.

(b) Injection moulding would suit chair A because its design requires a single piece of material with no additional parts. The flat thin surface of the design would not be possible with compression moulding or any other plastic moulding process so injection moulding was chosen.

(c) Polypropylene for chair A and ABS for chair B.

Why is this a top pupil answer?

(a) The marks available for this question would suggest that there is one mark available for each chair. The pupil has taken the time to write more than is required for the injection moulded chair and gives enough information to gain the mark available for the rotational moulded chair. (Full marks)

(b) This answer shows that the pupil recognises why the design of the chair lends itself towards injection moulding. Two reasons are given which is enough to gain the 2 marks available. (Full marks)

(c) One material is stated for each chair and they are different. (Full marks)

EXAM EXAMPLE 2

2 Name two features you would look for on a product to identify that it had been die cast. (2)

This is an average pupil answer.

I would look for a mark to show where the metal was poured in and marks to show how it was pushed out of the mould.

Why is this an average pupil answer?

This answer shows that the pupil knows a little about die casting and would probably be able to recognise a die cast product when shown one. However, the pupil does not use the correct terminology to describe what should be looked for and the answer may not attract full marks. (At least 1 mark would be awarded for this answer)

This is a top pupil answer.

Die casting metal is a process similar to injection moulding plastic in many ways. A product that has been die cast would show ejection marks. It would feel heavy. It may also be made in one piece even though it is a complicated shape. It would also be possible to see a seam line where the mould had been opened.

Why is this a top pupil answer?

The pupil has given more than two features that would identify that a product had been die cast. Again, this is always advisable especially if you find it difficult to explain or describe the examples. The correct technical vocabulary has been used. (Full marks)

EXAM EXAMPLE 3

3 (a) Why is it necessary to have tapered sides on a component that has been injection moulded? (1)
(b) Write down three other features that you would look for to identify that a product had been injection moulded. (3)
(c) Why is injection moulding only suitable for mass-produced products? (2)

This is a poor pupil answer.

(a) Tapered sides on the product allows it to be removed from the injection moulding machine easily.

(b) 1. An injection point 2. Ejection Pins 3. Ribs

(c) Because it is very expensive.

Why is this a poor pupil answer?

(a) The candidate is confused between the injection moulding machine and the mould for the component part. However, they do know that the taper is necessary for easy removal. Still the answer is not accurate enough for the one mark on offer. (0 marks)

(b) Ejection pin marks is the correct answer. Ribs are included for strength and are also part of products that have not been injection moulded. An injection point is correct. (1 mark would be awarded)

(c) What is very expensive? The pupil has made a broad statement and not backed it up. (No marks could be awarded)

This is a top pupil answer.

(a) Tapered sides are necessary on each component that is to be injection moulded so as it can be removed from the mould easily.

(b) Three other features that would identify that a component had been injection moulded are (i) Ejection pin marks (ii) an injection point mark (iii) a seam or line showing where the mould had closed.

(c) The initial set up costs are very expensive. Producing one mould for one part of any product will run into thousands of pounds. The injection moulding machine itself is also very expensive. Therefore to produce one product that has been injection moulded would cost several thousand pounds if not tens of thousands of pounds. However once everything has been set up the system is automated and can produce hundreds of thousands of parts for just the cost of the plastic which is very cheap. This means that the unit cost of each product would be very inexpensive.

Why is this a top pupil answer?

(a) The answer shows a clear understanding of why tapered sides are necessary. (1 mark)

(b) Each of the three features given is correct and has been well described. (Full marks)

(c) The pupil makes a good comparison between the cost of producing one product and the cost of producing thousands. This shows a good understanding of why injection moulding is only suitable for mass production. (Full marks)

EXAM EXAMPLE 4

4 The child's toy shown could either have been batch or mass produced.

(a) Describe the process of:

(i) batch production (1)

(ii) mass production (1)

(b) Describe how manufacturers use planning systems to organise production processes. (3)

(Higher Craft and Design, Question 4, 2002)

This is an average pupil answer.

(a) (i) Batch production means that quite a lot of the product is produced but not thousands and thousands which would be mass production. The process is best carried out in a traditional workshop with groups of people making different parts of the product before it is assembled.

(ii) Mass production is where thousands of the product is produced by machines.

(b) There are lots of people involved in making a product these days.

Manufacturers will use planning systems to help everyone know what they are doing. It is important that all the work is planned and everyone works together to finish the product on time.

Why is this an average pupil answer?

(a) (i) This is a reasonably good answer and would deserve the 1 mark on offer. (1 mark)

(ii) No description of the process has been given. This answer only makes an attempt to define what mass production is, it does not describe it. (0 marks)

(b) This is a very general response and not nearly enough information has been given for the 3 marks on offer. The pupil has not made any attempt to show any depth of knowledge on this topic. (1 mark)

This is a top pupil answer.

(a) (i) Batch production is a type of production that is best suited to smaller quantities that don't involve mass production techniques. Groups of workers will cooperate and share expertise during production. The parts for the product will be made in batches using jigs and templates to ensure that they are identical before being assembled.

(ii) Mass production usually requires machines and equipment which is automated. Vast quantities of identical parts can be produced using high tech equipment. It is a very repetitive process.

(b) There are various methods of planning production systems to make sure that they run smoothly and efficiently. Just in time production is one system that requires good stock control and efficient ordering of materials and parts. Sequential manufacturing is another best suited to assembly line manufacturing. This relies on nothing breaking down otherwise the assembly line is held up. Good maintenance and back up is required here. The other method is concurrent manufacturing which usually requires a Gantt chart to be produced. This divides the work up so that various parts can be manufactured at the same time and eventually brought together for assembly.

Why is this a top pupil answer?

(a) (i) Good information has been provided. The pupil mentions the use of jigs and templates, showing a deeper knowledge of this type of production. (Full marks)

(ii) The main points are correctly identified. The process is repetitive, automated and best suited to vast quantities of identical products being made. (Full marks)

(b) The pupil has described three methods of planning production and has made at least one good point about each. (Full marks)

9 MATERIALS

EXAM EXAMPLE 1

1 (a) Write down three advantages composite materials have over traditional materials. (3)
(b) Give an example of a product for each advantage. (3)

This is an average pupil answer.

(a) Composite materials are stronger, lighter and last longer.
(b) Furniture, sports equipment and kitchenware.

Why is this an average pupil answer?

(a) The three points that are made are true but need to be put into context by giving examples of how the composite material has replaced a traditional material. A comparison needs to be made to show that the pupil is fully aware of the improvement that the composite material offers. (It may be that 1 or 2 marks are given for this answer)

(b) The examples given are not specific. Actual products should be listed. The pupil should be able to draw on examples of products investigated during course work. (Probably 0 marks would be awarded here)

This is a top pupil answer.

(a) Composite materials can offer many advantages over traditional materials. An example would be a tennis racket. Traditional materials used in the past would have been solid timber. Now they are made from carbon fibre which is lighter and more durable giving obvious advantages to the user. This composite material can also be moulded much easier than solid timber which should offer ergonomic advantages over timber.

(b) Light weight – Racing car body shell
More durable – Racing ski for snow
Easier to mould – Jug kettles

Why is this a top pupil answer?

(a) A well explained answer that gives a comparison between the traditional material and the composite material and the advantages offered to the user. (Full marks)

(b) Three specific products are listed that are appropriate. (Full marks)

EXAM EXAMPLE 2

2 The kettle shown has been designed by Michael Graves for Alessi.

(a) Give two functional and two aesthetic reasons why stainless steel has been used for the main body of the kettle. (4)

(b) Why is polyamide a suitable material for the handle? (2)

This is a poor pupil answer.

(a) Stainless steel looks good. It is modern and fits in with a modern kitchen. It is easy to clean and does not rust.

(b) Polyamide is easy to manufacture.

Why is this a poor pupil answer?

(a) The statements that are made lack any evidence of depth or understanding. (1 or 2 marks could be awarded).

(b) This is true also of many other materials and it is not the reason that polyamide was used. (0 marks)

This is a top pupil answer.

(a) Functional reasons
Stainless steel is a hardwearing material that will withstand bumps and knocks in the kitchen. It is an easy material to clean and it is very durable.
Aesthetics reasons
Stainless steel has a very reflective surface which many people find attractive. It looks hygienic and clean and is suited well to a modern kitchen. Many other kitchen products are made from stainless steel and so a stainless steel kettle would match.

(b) Polyamide is a hardwearing material which will not scratch easily. The stainless steel will get very hot and will be dangerous to touch. This heat will not transfer to the polyamide handle which will stay cool.

Why is this a top pupil answer?

(a) Two good points are made. (Full marks)

(b) Three good points have been made here. Any two would give the full marks available. (Full marks)

EXAM EXAMPLE 3

3 Describe four qualities which plastics offer a product designer over traditional materials such as wood and metal.(4)

This is an average pupil answer

Plastic is much easier to shape and mould and it comes in bright colours. It is much cheaper and can be recycled.

Why is this an average pupil answer?

Four points are given but no description of the qualities offered over wood and metal. It is wrong to say that plastic is cheaper. Even if this were true it is not a quality of the material. It is true to say that plastic can be recycled and that this is an advantage. Metals can be recycled too and therefore this cannot be listed as an advantage over metal and wood. (1 or 2 marks could be awarded)

This is a top pupil answer.

(i) Plastics come in almost any colour. The material is self coloured and this offers a huge advantage over the traditional coloured finishes that would have to be applied to wood and metal. This is time consuming and often not as permanent as the coloured plastic.

(ii) Plastics can be easily moulded and formed and so this material lends itself to mass production better than metal and wood. It is cleaner to work with.

(iii) Plastics are water proof and weather resistant and don't need anything added to the surface to protect it.

(iv) Plastic offers a designer a material which is very strong but lightweight. This has advantages for many products such as car body shells and casings for hand held products.

Why is this a top pupil answer?

This answer has been well set out into four distinct sections making it easy to mark. A good description has been given for each of the four sections. (Full marks)

EXAM EXAMPLE 4

4 A plastic milk container is shown below.

(a) Describe the blow moulding process used to manufacture the plastic milk container. (3)

(b) State the name of a plastic that could be used to produce that container and give two properties of the chosen material that make it suitable. (3)

This is an average pupil answer.

(a) This process is where air is blown into a mould to form plastic. The mould is opened and the formed plastic is taken out.

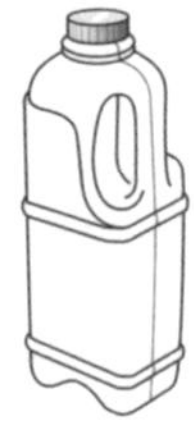

(b) ABS would be used to make the container. It is light weight and easy to form.

Why is this an average pupil answer?

(a) The pupil has not given a full description of the blow moulding process. There is nothing like the depth of knowledge required at Higher level. Perhaps 1 mark could be given for the pupil recognising that air is blown into a closed mould. (Possibly 1 mark could be given)

(b) ABS is not the most suitable material. The reasons given are also weak. ABS is not the most lightweight of plastics although it is easy to form. (Possibly 1 mark could be given)

This is a top pupil answer.

(a) A heated extruded length of plastic is placed in a closed mould. This plastic is soft and ready for moulding. Compressed air is forced into the mould and up through the extruded length of plastic forcing it out to the shape of the walls of the mould. The heated plastic is cooled by the cold walls of the mould and hardens to its shape. The mould is opened and the formed milk container is removed.

(b) LDPE could be used to make the milk container. It is a soft plastic which remains strong even when stretched as thin as the milk container. It is lightweight and suitable for the blow moulding process.

Why is this a top pupil answer?

(a) This is a good answer which is clear and easy to understand. It covers all the main points of the blow moulding process in sequence. (Full marks).

(b) A suitable material has been chosen and more than two properties which are suitable have been given. (Full marks)

REVISING FOR QUESTION 1

Question 1 in the written exam is worth 30 marks out of a total of 70 for the whole paper. This question has been written to test your knowledge of product design across a range of topics. It will give you one or two pictures of similar types of products and some information about them. You will then be asked several questions about these products. The questions are similar in nature each year and cover generally the same topics. These topics are likely to be:

- Writing a specification for one or both of the products in the pictures.
- Explaining why the materials used in the products are suitable for the production methods used.
- Explaining why the design of the product makes it suitable for the type of production methods used.
- Describing the appeal of the products to consumers.
- Describing the quality assurance issues surrounding the product.
- Explaining what health, safety and environmental issues should be considered during the design, production and use of the product.

Although the questions you will be asked will be similar to or the same as those of previous years, your answers will be different as they should be specific to the products shown and the information given about them. Therefore it is important that you have a good knowledge of these topics and are able to apply that knowledge to a wide range of products.

Before examining how to answer question 1 it would be useful to explore how you can build up your knowledge of such a wide range of products by doing the following exercises.

Revision exercise for question 1

Either individually or in a small group you should identify:

- 4 items of furniture
- 4 products that are used in the kitchen
- 4 products that are used in the home other than the kitchen
- 4 products that are used outdoors.

For each product you should provide a colour photograph and the following information on one A4 sheet of paper.

1. Write down the name of the product and state who you think is the intended market group.
2. Name the material used in the manufacture of each of the product's main parts.
3. Describe the features you have used to identify each material.
4. Justify the manufacturer's choice of materials for each part.
5. Name the production methods used for each part.
6. Describe what features you have used to identify the production methods used.
7. Why has this production method been used?
8. Write an outline specification for the product.

A sample answer for one product is shown below.

Product name	TV remote control by Panasonic
Market group	Household family use
Material	The main body is made from ABS.
Process	The main body has been injection moulded.
Identifying features of ABS	The inside of the battery cover lid has been stamped ABS. The material is light weight. The material is hard and scratch resistant. The material has a good quality surface finish.
Identifying features of injection moulding	Ejection marks inside the cover Injection point Tapered sides Radiused corners Strengthening webs Split lines
Justification for using ABS	It is a thermoplastic and is suitable for injection moulding. It is a durable hard wearing material suitable for 'heavy' use. It is suitable for mass production. It can be easily coloured before moulding. Cost effective as a mass production material.
Justification for choosing injection moulding	The process can produce the intricate shapes and surface finish required of the design of the main body. It is suitable as a mass production process. No further finishing required after moulding.
Outline specifications	Must be suitable as a hand held product. Should withstand being dropped regularly. Should be suitable for mass production. Should provide easy access to replace batteries. Should securely hold 2 AA batteries. Should provide easy finger access for all buttons required to operate the TV. Should completely and securely house the PCB inside the main casing. Should comply with colour and graphic conventions for function buttons. Should be low cost as a mass produced item. Should house the standard component (remote eye) in a suitable position to function when being held.

Now let us examine a typical question 1 and what you will be required to do on the day of your exam.

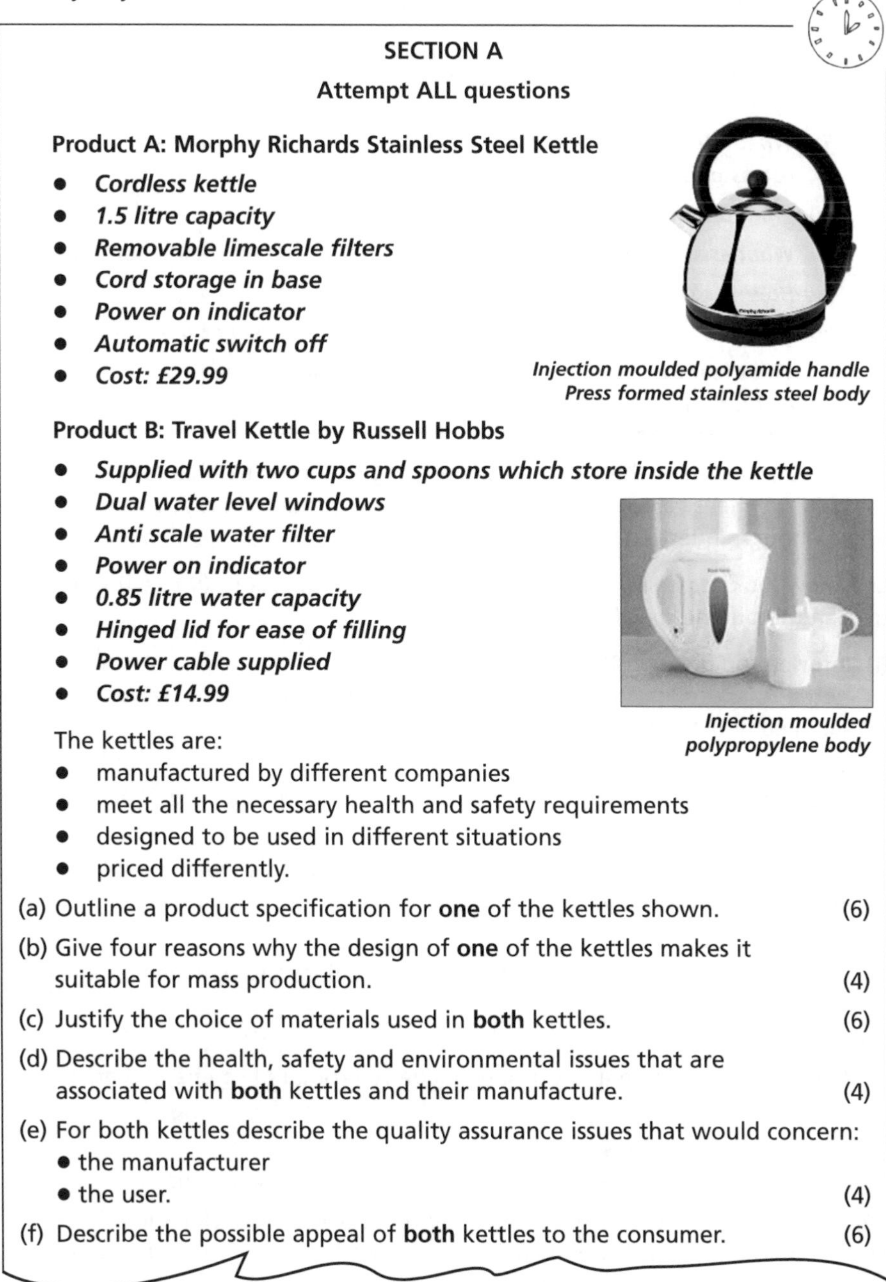

SECTION A

Attempt ALL questions

Product A: Morphy Richards Stainless Steel Kettle

- ***Cordless kettle***
- ***1.5 litre capacity***
- ***Removable limescale filters***
- ***Cord storage in base***
- ***Power on indicator***
- ***Automatic switch off***
- ***Cost: £29.99***

Injection moulded polyamide handle
Press formed stainless steel body

Product B: Travel Kettle by Russell Hobbs

- ***Supplied with two cups and spoons which store inside the kettle***
- ***Dual water level windows***
- ***Anti scale water filter***
- ***Power on indicator***
- ***0.85 litre water capacity***
- ***Hinged lid for ease of filling***
- ***Power cable supplied***
- ***Cost: £14.99***

Injection moulded polypropylene body

The kettles are:

- manufactured by different companies
- meet all the necessary health and safety requirements
- designed to be used in different situations
- priced differently.

(a) Outline a product specification for **one** of the kettles shown. (6)

(b) Give four reasons why the design of **one** of the kettles makes it suitable for mass production. (4)

(c) Justify the choice of materials used in **both** kettles. (6)

(d) Describe the health, safety and environmental issues that are associated with **both** kettles and their manufacture. (4)

(e) For both kettles describe the quality assurance issues that would concern:
- the manufacturer
- the user. (4)

(f) Describe the possible appeal of **both** kettles to the consumer. (6)

(a) Outline a product specification for one of the kettles shown.

Plan your answer

(i) ***What information has already been given in the question that I can use as part of my outline specification?***

(ii) ***What are the safety requirements of the product?***

(iii) ***What assumptions can I make about how the product should perform?***

(iv) ***How has ergonomics influenced the design of the product?***

(v) ***How has aesthetics been used to influence the appeal of the product?***

(vi) ***What features have been included in the product to improve its function?***

By examining each of these six areas carefully you should be able to write a fairly comprehensive outline specification for **any** product that you may be asked about in question 1.

Let's now look at possible answers to part (a) using questions (i) to (vi) to help us plan the outline specification.

(i) Information given in the question
Kettle A is cordless and has a 1.5 litre capacity. Kettle B has a power cable supplied and has a 0.85 litre capacity.

(ii) Safety
Electrical safety, stability, steam burns when refilling.

(iii) Performance
Time taken to boil, ability to pour accurately, method of opening lid, automatic power cut off.

(iv) Ergonomics
Comfort of handle, ease of pouring, ease of filling, gap between handle and main body.

(v) Aesthetics
Appeal to intended market group, image to promote hygiene and efficiency, corporate identity.

(vi) Features and functions
Power on indicator, water level indicator, automatic power cut off.

We have now made a plan of some of the things we could include in this outline specification and are ready to answer the question. There are 6 marks available for this part so a minimum of six points should be made but you should aim to write more. Remember to clearly state which product you are writing about.

(a) Product B outline specification

1. *This kettle has a maximum capacity of 0.85 litres to suit its purpose as a travel kettle.*
2. *This kettle is supplied with a power cord.*
3. *The electrical parts in the kettle should be fully insulated and protected from any water.*
4. *The kettle must be stable and not able to be knocked over easily with normal use especially when full of boiling water.*
5. *It should be possible to refill the kettle immediately after boiling without the risk of burning your hands from steam.*
6. *The kettle should pour water accurately into a cup without spilling it.*
7. *The handle of the kettle should be comfortable to hold to allow the user to use it safely and easily.*
8. *The gap between the handle and the main body of the kettle should be big enough for the 95th percentile male hand.*
9. *The kettle has a water level indicator to help prevent dry boiling and overfilling.*

Tips for success

- You will notice that not every point identified at the initial planning stage has been used. Some have been left out because there is a doubt as to whether it may be true for that product, e.g. does Kettle B have an automatic power cut off? While this may be desirable it is not something that is present in every kettle.
- The information given is thorough with a reason being given to justify or explain its inclusion. For example, point 3 does not simply say that the kettle should be stable, it gives some indication of the level of stability expected and why stability is important. This additional information will show your level of understanding about the product's requirements.

(b) Give four reasons why the design of one of the kettles makes it suitable for mass production.

Plan your answer

(i) Will the materials used to make the product impose a restriction on the choice of production methods that can be used?

(ii) Will the complexity of the shape of the parts of the product determine that a certain production method is used?

(iii) Will the type of construction, joining and assembly methods used make the product suitable for a particular type of production?

(iv) What quantity is to be produced?

(v) Has a stock size of material been used?

(vi) Has the same part been used more than once in the product?

Other factors that may influence the type of production that can be used are time, money, the expertise of the workforce and availability of equipment.

Now let's look at a model answer for part (b). Remember to state the product you are referring to.

(b) Kettle A

1. *The kettle has been made from stainless steel which is a material suitable for high volume mass production.*
2. *Consumers will expect a high quality finish on a commercially produced product such as this one and this is best achieved through press forming which is a mass production process.*
3. *The holes in the stainless steel main body for the lid, spout and base need to be accurate and aligned properly for each of the thousands of kettles to be manufactured. Therefore setting up a mass manufacturing production line to pierce these parts prior to press forming the main body will reduce the possibility of human error.*
4. *The complexity of the shape of the handle and the thermoplastic material (polyamide) chosen make it suitable for mass manufacturing through injection moulding.*

Tips for success

- You will notice that in point 4 the answer shows that the candidate is aware that the shape of the handle is complex and that polyamide is a thermoplastic. Both of these points justify injection moulding, which is a mass manufacturing process, as the chosen method of production.
- Try to give some information to justify the point you are making. You will notice in 2 that the consumer expectations of the quality of surface finish for this mass produced product reinforces the need for press forming.

(c) Justify the choice of materials used in both kettles.

Plan your answer

(i) In what type of environment is the product expected to be used?

(ii) Consider the safety of the person who uses the product.

(iii) How is the product expected to perform? (weight, stability, strength, etc.)

(iv) Consider the appearance of the product to its intended market group.

(v) What type of manufacturing processes have been used to make the product?

(vi) Is the material able to be recycled or reused?

Now let's look at a model answer for part (c). Remember to state the product you are referring to.

(c) Kettle A

1. *Stainless steel is a good choice of material for this kettle because it does not corrode which is vital for a metal that is in contact with water.*
2. *Stainless steel has the ability to keep its shiny reflective appearance allowing the product to maintain its aesthetic qualities throughout its life time.*
3. *Polyamide is a strong scratch resistant material which is useful as a handle as it does not conduct heat. It can also be moulded much easier than stainless steel to the required shape.*

Kettle B

1. *Polypropylene is a good choice for this kettle as it can be injection moulded which is necessary given the intricate and complex appearance of the kettle.*
2. *Polypropylene is suitable for the bright white glossy surface of the kettle. It can retain these aesthetic qualities after the moulding process. The material in this form also gives the product a clean, hygienic appearance suitable for a kitchen.*
3. *Polypropylene is a lightweight material which is an advantage especially when the kettle is full. Its lightweight properties are also necessary as it is a travel kettle.*

Tips for success

- If you are asked to justify something you should give a reason which supports your claim that what you state is true. For example, for Kettle A point 1 supports the need for the material not to corrode by stating that the product will come in contact with water.
- There is no need to restrict yourself to giving only one reason for each answer. For example, in Kettle B point 2 the answer touches on the material being suitable for injection moulding as well as its aesthetic qualities.

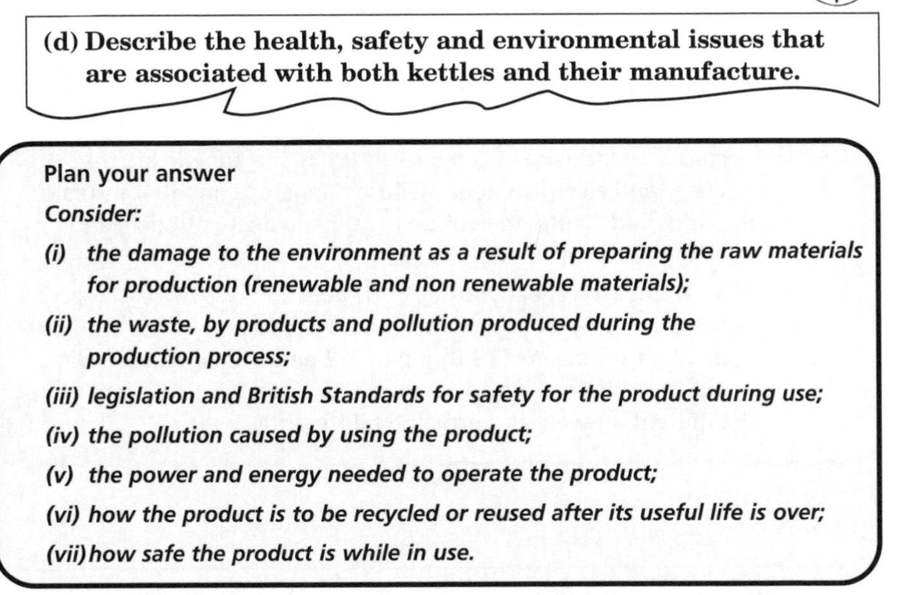

(d) Describe the health, safety and environmental issues that are associated with both kettles and their manufacture.

Plan your answer

Consider:

(i) ***the damage to the environment as a result of preparing the raw materials for production (renewable and non renewable materials);***

(ii) ***the waste, by products and pollution produced during the production process;***

(iii) ***legislation and British Standards for safety for the product during use;***

(iv) ***the pollution caused by using the product;***

(v) ***the power and energy needed to operate the product;***

(vi) ***how the product is to be recycled or reused after its useful life is over;***

(vii) ***how safe the product is while in use.***

Now let's look at a model answer for part (d). This question has asked that both kettles are referred to.

1. *The most obvious safety issue associated with both kettles is electrical safety given that they will come into regular contact with water. This will require that all electrical components are fully insulated.*
2. *Both kettles have moulded plastic parts. It is important these parts have been labelled clearly so as they can be identified for recycling.*
3. *During manufacture it is important that any waste is disposed of in accordance with required legislation and that excess material is recycled and reused.*
4. *Both kettles are electrical. In order to conserve energy in the home consideration should have been given to how much electricity is required to boil the kettle and that both kettles are capable of boiling just enough water for one cup without boiling dry or damaging the electrical element.*
5. *The stability of the kettle is important. It should be designed so as not to topple over easily especially when full of boiling water.*

Tips for success

- Try to back up your claim by giving a reason which in your opinion justifies the point you are making. For example, in point 1 it is not enough to say that electrical safety should be considered; this has been backed up by the knowledge that water and electricity together are dangerous and that insulation is required. These obvious points show a deeper understanding of the problem.
- Try not to repeat yourself. This question was worth 4 marks, therefore at least four points should be made. Make a different point each time. Notice that points 2 and 3 deal with recycling but in a different way. Therefore they have been separated into different answers and explained differently.

(e) For both kettles describe the quality assurance issues that would concern:
- **the manufacturer**
- **the user.**

Plan your answer

The manufacturer will be concerned with the:

(i) quality of the raw materials delivered;

(ii) prompt and efficient delivery of materials (Just In Time Production);

(iii) accuracy and precision of each part prior to assembly;

(iv) quality and performance of the product after assembly;

(v) efficiency of the production team to meet deadlines;

(vi) performance of the finished product and how well it meets health and safety legislation.

The consumer will be concerned with the:

(i) safety of the product and its ease of use;

(ii) effect the product will have on the environment;

(iii) product's ability to perform to the same standard over time;

(iv) ability for the product to be repaired and maintained;

(v) guarantees and warranties issued with the product.

These issues will remain the same for every product. You should write about them in terms of the product shown in question 1.

Now let's look at a model answer for part (e)

Quality assurance issues that will affect the manufacturer.

1. *It is likely that the different parts of the kettle will be made separately and come from a variety of different sources. Therefore it is vital to the efficient production and assembly of the kettle that these parts are delivered on time and are of a good quality, otherwise it is unlikely that the manufacturer will be able to meet deadlines.*
2. *The manufacturer will want to test the finished kettle to ensure that it meets health and safety standards. In particular it would be important that the seals and insulation are effective in keeping water away from the electrical components.*

Quality assurance issues that will affect the user

1. *The consumer will be very concerned that the kettle meets all the required electrical safety standards and is reliable throughout its usable life. They will want assurances that the kettle is reliable and safe and that this is backed up with a guarantee.*
2. *Consumers are now more environmentally conscious and may take an interest in how energy efficient the kettle is. For example, what is the kettle's power rating and is it able to boil only a cupful of water at a time?*

Tips for success

- The points raised in the planning box are common to most products. Make sure you write your answer specifically about how these issues will affect the product you have been asked about.
- Similar issues will affect the user and the manufacturer but in different ways. Notice that the user will be affected by the long term electrical safety and that the manufacturer may only concern themselves with the immediate quality assurance of a new product.

(f) Describe the possible appeal of both kettles to the consumer.

Plan your answer

Consider the product's:

(i) aesthetic appeal to the intended market group;

(ii) safety features;

(iii) build quality and reliability;

(iv) brand name and the reputation of the manufacturer;

(v) cost compared to other similar products;

(vi) features and functions and what benefits they provide.

Remember to describe the appeal of both kettles and be sure to name the one you are talking about.

Kettle A would appeal to many consumers. The stainless steel body gives it an attractive modern image that would not look out of place in most kitchens today. The black handle and base fit well with this material and provide it with a mature classic look. This would not be achieved if a bright colour which dominated the product had been chosen. Morphy Richards is a well known manufacturer of electrical goods which may give the consumer confidence in the product. The product has a range of features that make it convenient and easy to use.

Kettle B is a small travel kettle that travellers and holiday makers may find useful. Its size may also appeal to people living alone or to hotels and guest houses for use in their rooms. It is relatively cheap compared to bigger domestic kettles which appeal to consumers especially if it was to be used infrequently. The inclusion of cups and spoons would suggest that it would be particularly useful to the traveller especially as they can store inside the kettle.

Tips for success

- When you make a claim that a product would appeal to a consumer try to support the claim with a reason. For example, it is true that Kettle A has a range of features but its appeal is that consumers will find that these features make the kettle more 'convenient and easy to use'.
- If possible try to write about different things for each product.

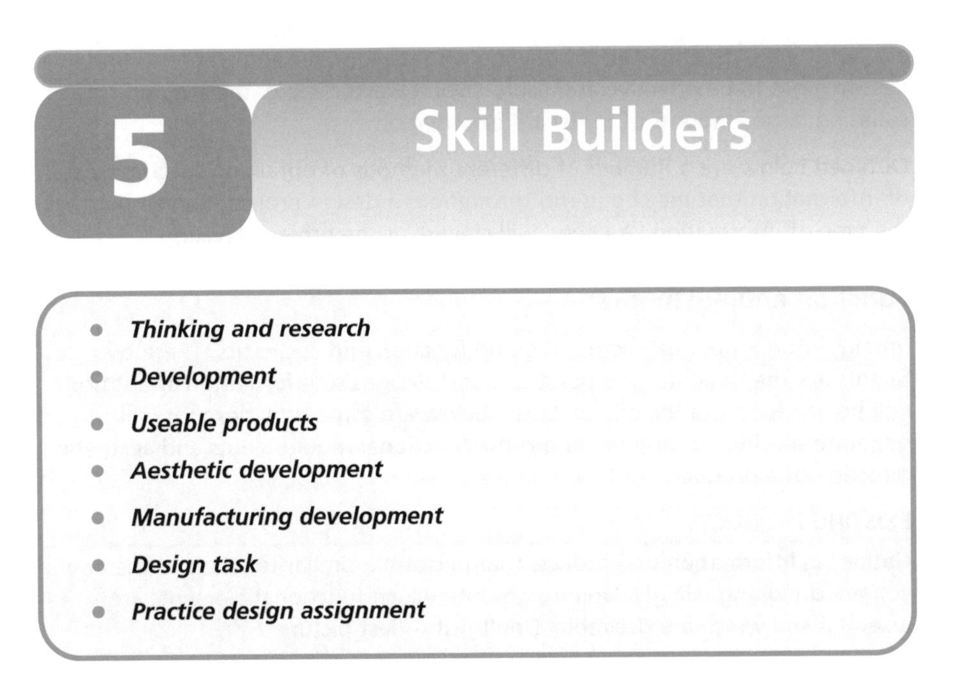

This chapter contains exercises that are intended to supplement the skills you will already have gained from completing the Higher units. It is in no way comprehensive in its treatment of the design process, but will hopefully help in your preparation and execution of the design assignment. Drawing on your previous skills, knowledge and understanding, the activities have been developed under the broad headings of:

- thinking and research
- development and communication
- evaluation and presentation.

THINKING AND RESEARCH

As mentioned in the previous chapter, design is a thinking activity, based on learned skills rather than just intuition. Asking the right questions and making informed choices are essential for successful design work to take place. Research is the key to doing well in this subject. Thinking of design in terms of improvement rather than invention makes the design process more achievable for pupils of all abilities. Most of the problems you come across or decisions you have to make will have been encountered before, by someone else. The answers are out there – it's a matter of knowing what to look for and where to find it. Research has to be effective and useful. Don't waste time gathering and collating large amounts of irrelevant information.

Outlined below are a number of different methods of obtaining specific types of information that may be useful throughout a design project. Remember that the type of information you need will change as the project develops.

Function and aesthetics

Initally, information may be required on function and aesthetics. These two factors are the basic starting point for most designs so information about both will be needed from the outset. Listed below are some activities that will generate specific information about the functional requirements and aesthetic direction of a product.

EXISTING PRODUCTS

Gathering information on products that perform a similar function to the one you are designing will give you insight into which functional elements are essential and which are desirable. Don't just collect pictures.

Task

Gather information about two irons, one from the top end of the market and one from the bottom. Use trade literature, magazines and the internet for this activity. Use the information to identify functions and features that are the same or similar. These are likely to be the functions and features that are fundamental to the success of the product. Try to identify which features have contributed to one iron being a top of the range model. Essential elements should be incorporated in any concept; other additional features may or may not be included depending on the product's intended purpose.

LIKE PRODUCTS

Looking at products that are similar often identifies weaknesses, or highlights a need for improvement. Rather than beginning from scratch you could identify products that have attributes that could be incorporated into the design of your product.

Task

Use books, magazines and the internet to identify products that perform in the same way as a household vacuum cleaner. When gathering this information consider how the product functions, how it works and how it is used.

LIFESTYLE BOARDS

The look of a product will usually be heavily influenced by its intended user group. Gathering pictorial information about what appeals to them on topics such as hobbies, interests, socialising, etc., can provide a visual stimulus for shape, form, colour, materials and overall styling.

Task

Produce two lifestyle boards, using quality images. Remember that the purpose of this activity is to generate visual impact. One lifestyle board should be centred around a typical teenager's lifestyle and the other for the over 60s. Identify the different styles, forms, colours, etc., that make them visually different.

Ergonomics and safety

These two factors will usually begin to become more important during the developmental stages of the design process. Developing a design could be viewed as 'getting real', and making sure that it is user friendly and safe will play a major part in this. How can you identify what is important?

USER TRIP

This activity can be used to gain information, some of which can be taken for granted, such as interactions required for the correct use of a product. Try to imagine that it is the first time you have used the product when doing this activity.

Task

Write down all the interactions required to make a cup of tea using a household kettle. Now make a cup of tea using your own kettle, noting down all the times you interact with the kettle and how you interact with it. The list is likely to be much longer and should generate more insight.

ANTHROPOMETRIC DATA

This information is essential when working to scale. Determining the correct percentile will come from thorough analysis and understanding of the use of the product.

Task

Using the information gathered from the user trip, identify what anthropometric data would be required when designing the kettle. Determine whether this information should be taken from the 5th, 50th or 95th percentile.

SITUATIONAL ANALYSIS

Analysing a product's intended use, the user and its environment, together with its potential for misuse, will generate useful information. Real safety issues will be highlighted that should be addressed in the product's development.

Task

Write down all the opportunities for causing injury or harm that can arise when making a cup of tea. Don't take things for granted, do the activity yourself. When carrying out this activity consider other user groups who might be expected to use the product, such as the elderly or younger children.

Materials and construction

Materials and construction will become increasingly important as the design develops, but should be considered throughout the process as they offer interesting development opportunities, and impact on all elements of a product's design.

EXISTING PRODUCTS

Looking at existing products will often provide information on suitable materials and construction methods, providing insight into why particular materials and construction methods are used.

Task

Using your own household toaster determine what materials were used in its construction and why they were considered appropriate in terms of cost, manufacture, function and aesthetics.

PERFORMANCE ANALYSIS

This can often highlight the most essential information required to ensure that a product's construction and materials are suited to its intended purpose.

Task

Highlight the performance required from the different parts of your own household iron, i.e. the handle, body and sole. Write down how the performance of each part has influenced the choice of material.

DEVELOPMENT

Familiarity makes design difficult

Before you begin to design a product it may be useful to identify and understand its requirements. This will provide a good foundation on which to develop the design. Unfortunately this is not always an easy task. Everyday products are taken for granted, and largely go unquestioned. This can lead to poor superficial redesign. For example, the typical image of a table tends to be a flat rectangular surface with four legs.

That image is often the starting point for most table design, but may not be the best one if greater understanding is required. The example below offers a more interesting starting point from which to begin a table design. How to hold up the surface presents an interesting challenge, more conducive to idea generation.

TASK

- Don't think table, think surface.
- How is it held up?
- How could it be held up?
- What could I hold it up with?

Use sketches similar to those shown to explore table design. Keep the proportions constrained to the rectangle as this will make sketching in both 2 and 3D easier.

Tip: Add more constraints to generate more creativity. Try restricting the material you use – can you hold the surface using only: MDF, sheet steel, steel tubing, acrylic? How does this affect the construction of the table?

USEABLE PRODUCTS

Developing a basic solution may help stimulate creative thought, but will not necessarily provide a successful product. For a design to be successful it must be suitable for its purpose. It will usually have a number of key features that allow it to perform its intended task well.

The key functions and features of a design are usually heavily influenced by the user or target market together with its intended environment.

TASK

Assume that the basic function of a seat is to support a human body in a seated position. This is shown in the picture below in its most basic form – a suspended surface and back rest.

As with the table, a means of support will have to be developed for the seat. But at the same time consideration will have to be given to the seat's user.

Consider how the features of the seat would differ if the user was to be:

1. a toddler
2. a school pupil
3. a dentist
4. an umpire.

Select a user from the list above or think of one yourself and, keeping the environment in mind, use the type of sketches shown opposite to develop a functional design. Try to ignore the aesthetic of the seat at the moment; it is likely that experimenting with support, construction and function will create an interesting look in the end.

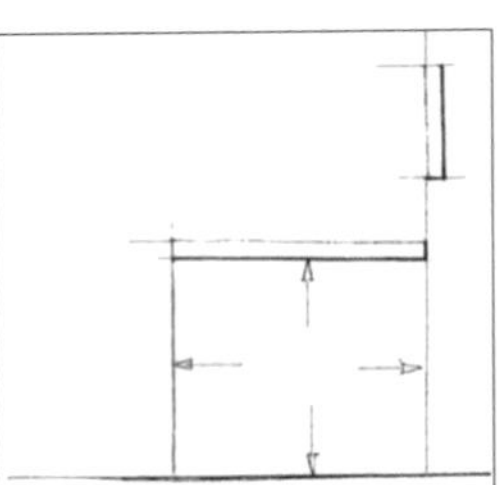

Tip: Try to build your design up, adding to it slowly. Start by supporting the horizontal surface, then add features one at a time and evaluate their impact. What anthropometric data would be relevant?

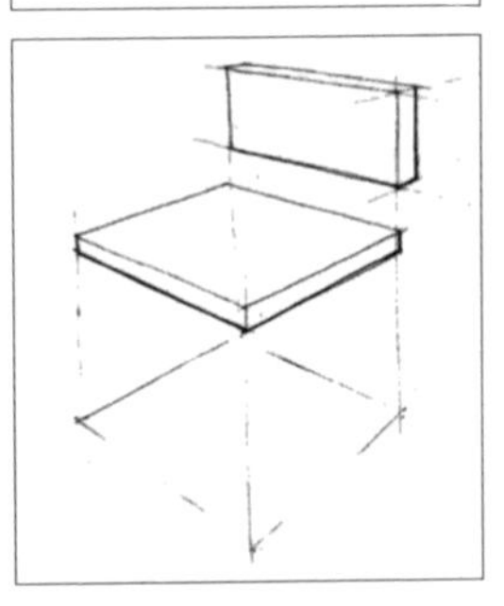

AESTHETIC DEVELOPMENT

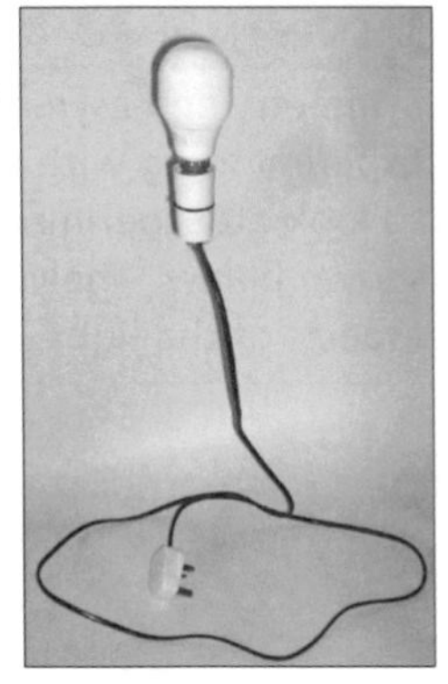

The photograph opposite shows the components necessary for a light. Fixings, knockdown fittings or complex PCBs can offer a good starting point for a design, but its appeal and ultimate success will be determined by its look.

Standard components such as these can offer an interesting starting point for a re-design, allowing the aesthetic to evolve naturally.

Before using mood and lifestyle boards to generate an aesthetic direction, it is useful to experiment and analyse some of the basics. This will build up a visual understanding and vocabulary required throughout development. Words such as 'proportion', 'symmetry', 'balance', 'contrast' and 'colour' are often used to describe a product's aesthetic. It is important therefore that you can both understand and use them knowledgeably.

TASK

Proportion can be overlooked when developing a design proposal. By altering the proportion of a product you can change its appearance from squat and heavy to light and elegant. Experiment with proportion. Sketch the design for a light. Evaluate how stretching and squashing the base impacts on the design's appearance.

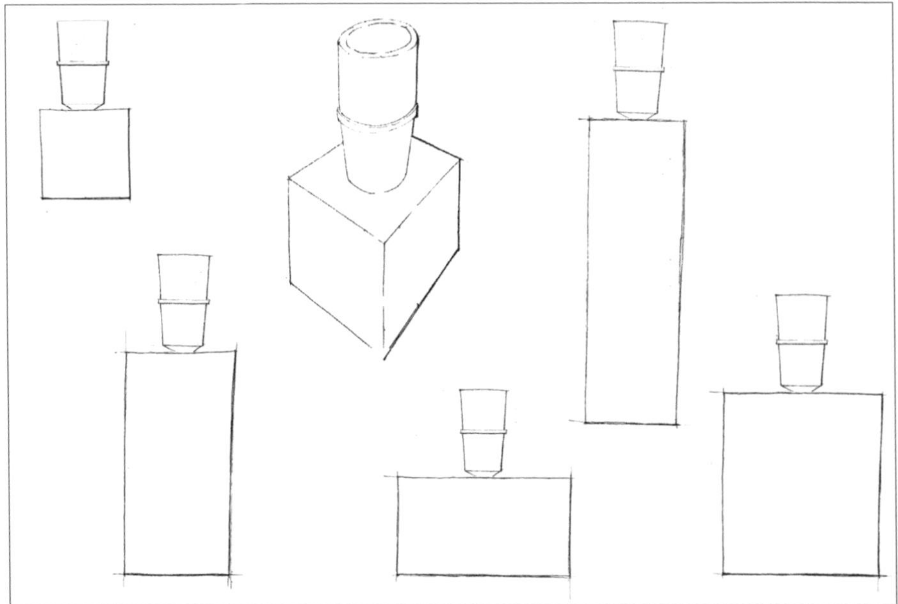

TASK

Symmetry and asymmetry can contribute to the visual impact of a design. Experimenting with visual balance can create a product that is more interesting to look at. Experiment with the symmetry of the light using the type of sketches shown below. Analyse the impact this has on both the aesthetic and functional aspects of the light.

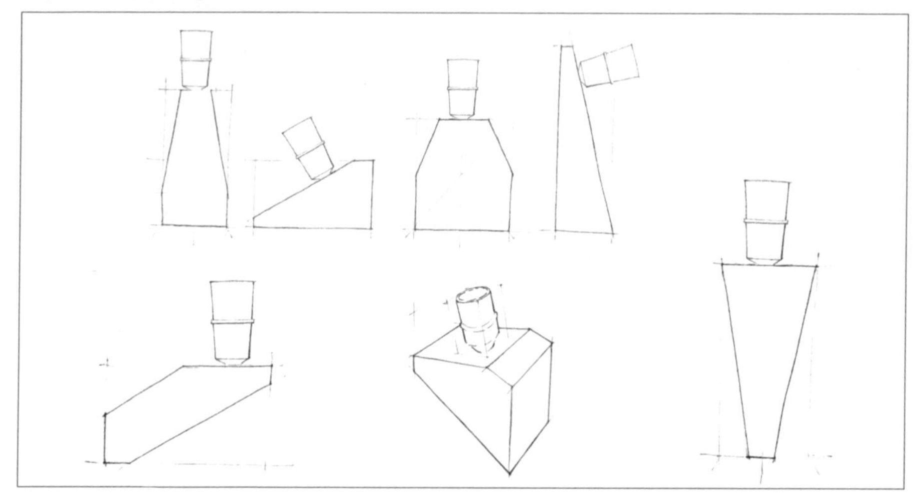

TASK

Balance can be used to create tension and interest in a design. Ignoring the effects of gravity and stability for the moment, use the type of sketches shown below to experiment further with a light which is visually unbalanced.

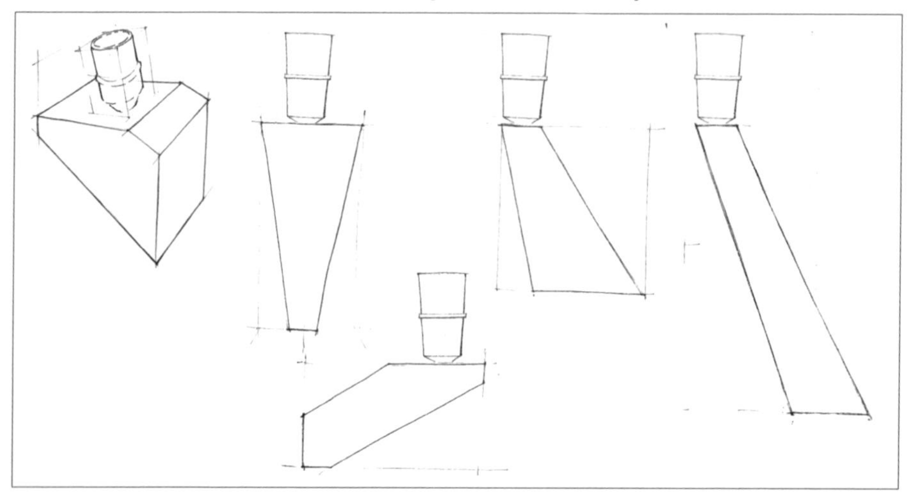

TASK

Harmony and contrast offer a range of opportunities for aesthetic development, involving shape, form, colour and texture. Develop your design further by adding a base which may or may not be in harmony with your previous design work for the light. You may have to introduce rendering techniques to highlight material, colour or texture. Identify through research a range of existing shades for the light you have produced. Consider harmony and contrast here as well.

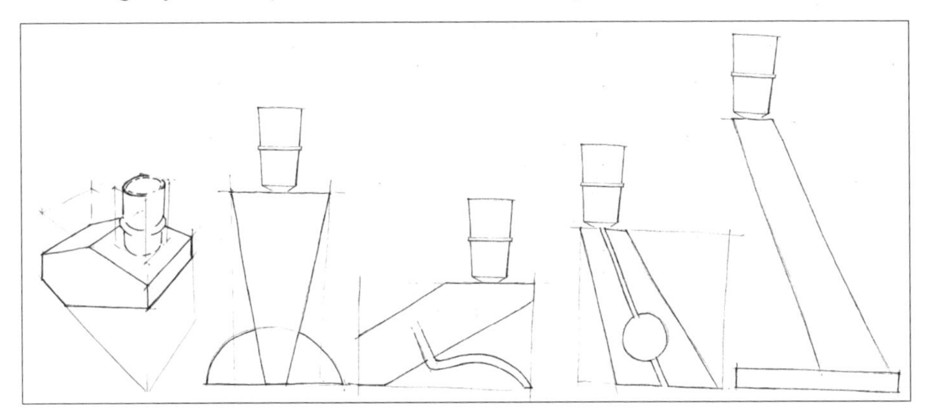

MANUFACTURING DEVELOPMENT

Design does not always require radical change. A more focused development confined to one element of the design process often creates a more thoughtful response that is manageable and easy to evaluate.

Below are two photographs detailing a pupil's design for a pencil holder. It was manufactured as a one off, reliant on high levels of craft skills which would make it uneconomical to produce commercially.

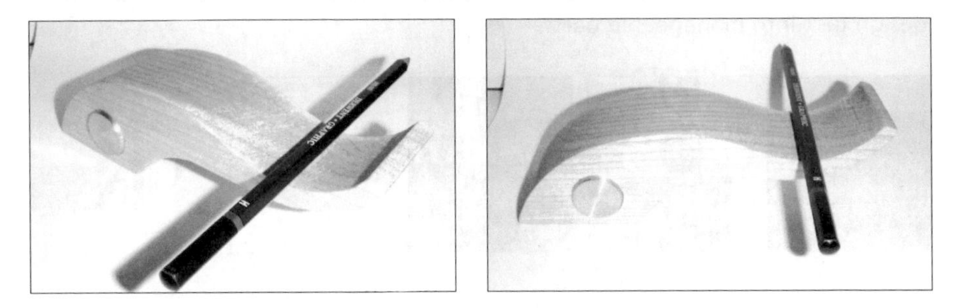

TASK

Using 2 and 3D sketches, sections, details and annotation, explain the possible changes and features required in order to mass produce this pencil holder using the following processes:

Injection moulding

- Consider: split lines, wall thicknesses, ribs and webs.
- Explain the advantages and disadvantages of this process in relation to this product.
- Identify and justify suitable materials for this product using this process.

Extrusion

- Consider: sections, profiles, wall thickness and die shape.
- Explain how the form of the product may have to change slightly to be manufactured in this way.
- Identify and justify suitable materials, other than plastic, for this product using this process.

Rotational moulding

- Consider: concave and sharp sections, and thin weak areas.
- Explain how the problem of the counter balance could be addressed.
- Identify and justify suitable materials for this product using this process.

DESIGN TASK

The design task below will provide the opportunity to bring together the skills outlined in this chapter. Read the design brief carefully, looking for clues and possible avenues for further research. Use the research provided and break the design task into manageable parts.

Design brief: *Disposable Camera Project*

The world of fashion is a highly competitive field. It is image conscious and heavily reliant on branding. In order to maintain a high profile, new and innovative advertising campaigns are constantly required.

You have been selected to develop a disposable 35mm camera to be used as part of a new advertising campaign. The camera is to be given away at the point of sales, to customers who spend more than £50 in any retail outlet.

The camera campaign will be used to promote a fashion house, enhancing their reputation in the high street for quality and cutting edge design.

The camera, although a throw-away item, will have to retain a quality and image comparable to the company's other products, taking care to maintain consumers' high expectations of the brand.

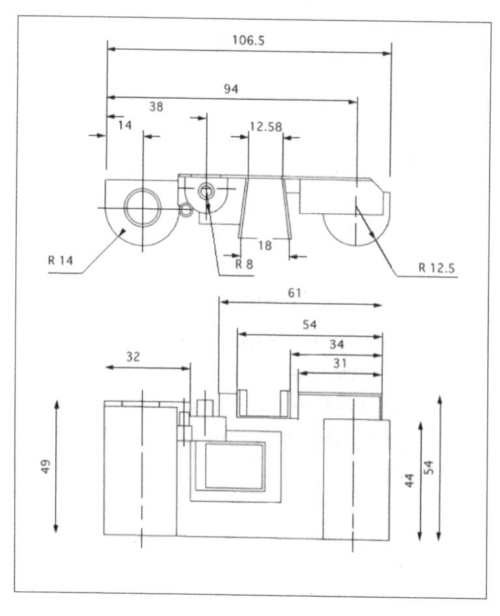

Standard component: Internal chasse with dimensions

Use the photographs to highlight design features that make it functional, disposable, cheap to produce and manufacture.

Use the existing product to identify problems with the product. Why would it not satisfy the design brief?

TASK 1

Research a fashion house/designer, including examples of the company's product range, typical retail outlets; together with a consumer profile or target market. Work could be presented in the form of:

- mood boards
- lifestyle boards
- product boards.

TASK 2

Analyse an existing disposable camera; it may be beneficial to consider the following for your analysis:

- suited to its purpose
- ergonomics
- manufacturing processes
- materials
- assembly
- standard components

This work should be presented as a series of annotated sketches and photographs used to highlight the important factors in the product's design and development.

TASK 3

Develop a design proposal for the new camera.

- Provide a detailed specification to include performance, technical and marketing information.
- Provide initial ideas created through sketching and modelling.
- Develop a design that meets your own specification. Remember relevance to the design brief is essential. Ensure that aesthetics have been derived from the company's image and your mood board; and the design is user friendly, functional and suitable for cheap mass production. The camera must be based on the standard components provided.
- Record in writing the main decisions made throughout the design's development.
- Scale drawing 1:1.
- Construction and production information.
- Presentation drawing.

PRACTICE DESIGN ASSIGNMENT

Below is a design assignment, similar in nature to the ones set by the SQA. It includes some tips for success, which will not be given in the actual Design Assignment.

Design situation

The number of tourists visiting Scotland continues to grow each year. They are drawn to the country by its *complex history and beautiful scenery*. The most popular attractions tend to be castles, historic houses, battlegrounds, ruins and museums. However some clever *niche marketing* has been used to promote city breaks to Scotland's major cities. The capital city of Edinburgh, together with Glasgow, is becoming increasingly cosmopolitan, both destinations being seen as attractive alternatives to the countryside. This boosts the Scottish economy by creating *all year round tourism*.

The popularity of Scotland as a tourist destination has created pressure on local authorities to improve a number of *facilities offered to tourists and visitors*. All public spaces require special consideration. The overall appearance, attention to detail and quality will say a lot about the country's standards and attitudes.

Scotland enjoys an international reputation for *architecture, engineering and invention*. This should in some way be *reflected in the quality of its public spaces*. This civic redesign programme will involve a range of products that will improve the use and appearance of both urban and rural areas. All proposals should have a *strong Scottish aesthetic with a modern perspective*.

Three possible exterior furniture projects have been identified that could be used to enhance a wide range of public spaces. All the designs will have to be *flexible* and *adaptable* to suit a range of different urban settings. Particular attention should be given to durability and maintenance, as designs of this nature are often *targeted for abuse and vandalism*.

Initially the products will be limited to small runs to gauge public opinion. However, if successful, greater numbers will be required. This should be reflected in the design proposal.

Your task is to produce a design proposal for one of the three design specifications shown on pages 149–151. Your design proposal must meet all the requirements of the specification and should make use of the research material.

Research information

Possible urban environments

Edinburgh *Aberdeen* *Inverness* *Glasgow*

Tip: Select one of the areas from the four above. Use this to determine what design task you are going to attempt. Note down why the area would benefit from your design, together with any special features or requirements. It is likely that the environment will have a large impact on the design's aesthetic.
More information may be useful. Information on factors such as safety, durability, maintenance and materials may be discovered from a general analysis, using various lateral thinking techniques.

Anthropometric data table

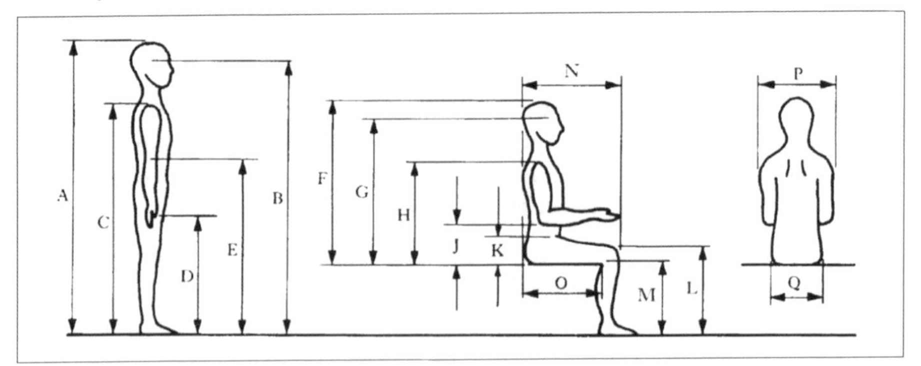

	MEN	WOMEN	MEN	WOMEN	MEN	WOMEN
DIMENSIONS	5%le	5%le	50%le	50%le	95%le	95%le
A Stature	1682	1523	1740	1610	1886	1728
B Eye Height	1544	1430	1625	1490	1742	1628
C Shoulder Height	1365	1229	1415	1310	1557	1414
D Knuckle			765	700		
E Elbow	105	965	1080	980	121	109
F Sitting	880	810	910	845	995	910
G Sitting Eye	764	695	800	740	865	796
H Sitting Shoulder	573	542	600	565	696	631
J Elbow Rest	188	180	240	225	295	279
K Thigh Clearance*	145	104	150	140	191	149
L Knee Height	521	467	540	496	603	543
M Popliteal Height*	393	356	440	415	490	445
N Buttock Knee Length	541	518	595	570	640	625
O Buttock Popliteal Length	439	432	495	480	549	533
P Shoulder Breadth	444	386	460	415	529	468
Q Hip Breadth	310	312	360	375	404	434

%le = percentile

Allowances of 25 mm for men and 45 mm for women must be made where indicated as measurements are for the unclothed person.

Tip: Analyse existing products that do a similar job to the product set out in your chosen brief and identify important interactions required from your design. Highlight the percentile ranges appropriate to your design. Make reference to these during the development stages. Include a scale drawing of your proposal using the data above. This is the only way to describe your proposal's ergonomics.

Standard bin insert

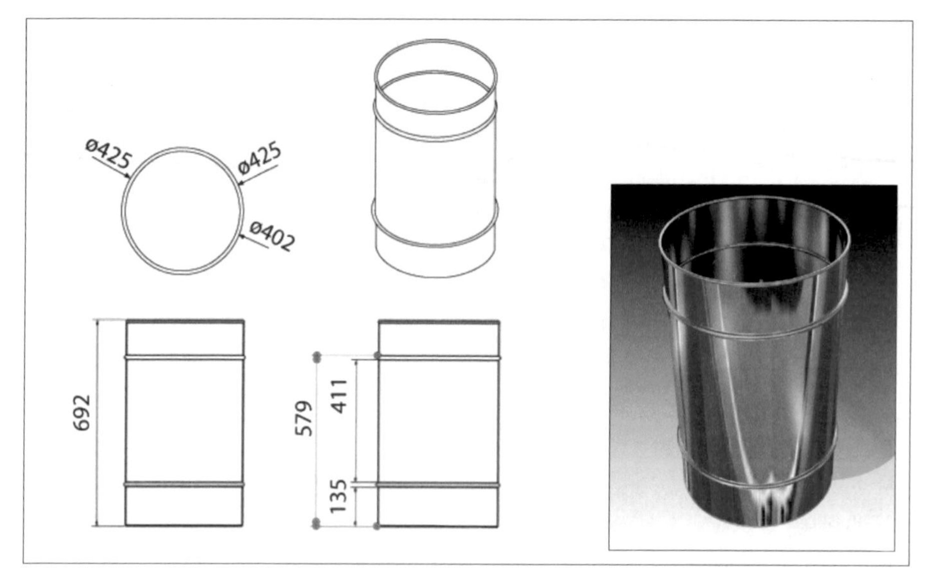

Touch screen display

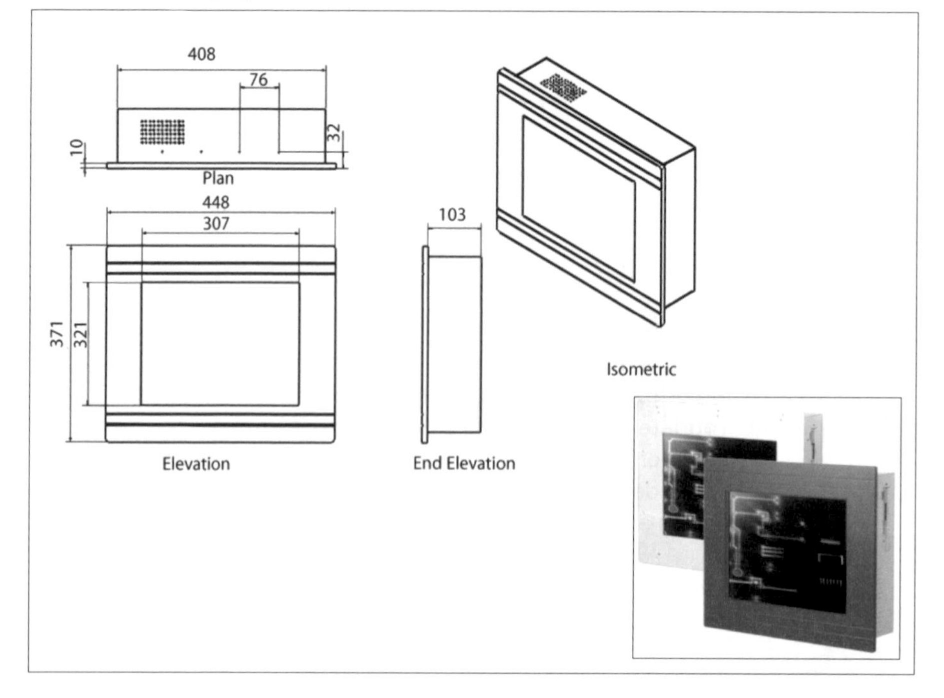

Tip: The components opposite should be incorporated into your design. They will influence not only the size and shape of the design but also some of the constructional detailing. You should show how you have understood and worked within the limits of these standard components, from initial ideas through to your final proposal.

Three design tasks

Choose one of the design tasks specified below.

DESIGN TASK 1: EXTERIOR PUBLIC SEATING

An exterior public seating system is required that can be used in a variety of urban public spaces. The design must be suited to a wide range of environments and provide a comfortable resting place for the general public.

Function

- The seating will provide comfortable seating for two adults.
- The seating will be of a modular nature, allowing greater flexibility.
- The seating will provide a method of fixing it in place.
- The seating will be resistant to the changeable Scottish weather.

Safety

- The seating will pose no threat to public safety.
- The seating must be safe and be able to withstand a degree of misuse/abuse.
- The seating construction must be robust and resistant to deliberate acts of vandalism.

Aesthetics

- The seating must create a strong Scottish identity, but avoid being kitsch.
- Designs are encouraged to utilise traditional Scottish materials if at all possible.
- Designs are to reflect a modern forward thinking Scotland.

Maintenance

- The seating system will provide easy maintenance and cleaning.
- The seating will be durable, lasting for a period of five years.
- Replacement and repair should be considered in the proposal.

Cost

- Initially 100 units will be required. However, the potential to mass produce should be considered.
- The unit cost for each bench should not exceed £400.

DESIGN TASK 2: EXTERIOR LITTER BIN

A design is required to encourage a more responsible attitude towards littering the environment. A new design of litter bins will be used to highlight the recurring nuisance of littering public spaces.

Function

- The litter bin will allow for easy disposal of rubbish.
- The litter bin will prevent and deter over filling.
- The litter bin will be resistant to the elements.
- The litter bin will provide secure access for refuse collection.

Safety

- The litter bin will pose no threat to public safety.
- The litter bin must be safe even through misuse.
- The litter bin's construction must be robust and resistant to deliberate acts of vandalism.
- The litter bin will be fire resistant.

Aesthetics

- The litter bin's design must create a strong Scottish identity, but avoid being kitsch.
- Designs are encouraged to utilise traditional Scottish materials if at all possible.
- Designs are to reflect a modern forward thinking Scotland.

Maintenance

- The litter bin will provide easy maintenance and cleaning.
- The litter bin will be durable, lasting for a period of five years.
- Replacement and repair should be considered in the proposal.

Cost

- Initially 200 units will be required. However, the potential to mass produce should be considered.
- The unit cost for each bin should not exceed £100.

DESIGN TASK 3: INFORMATION HUB

A display is required to provide information to visitors and the general public. It is to utilise modern touch screen technology, offering flexibility and an ability to change. It is hoped that these hubs will become landmarks and meeting points.

Function

- The information hub will utilise easy to read interactive displays.
- The information hub will allow two people to access the information simultaneously.
- Information must be accessible for all potential users.
- The information hub will incorporate the i symbol.
- The information hub will be permanently fixed to the ground.
- The information hub will be shut down overnight.

Safety

- The information hub will pose no threat to public safety.
- The information hub must be safe even through misuse.
- The information hub's construction must be robust and resistant to deliberate acts of vandalism.
- The information hub will be compliant with all British standards concerning electrical installations.

Aesthetics

- The design must create a strong Scottish identity, but avoid being kitsch.
- Designs are encouraged to utilise traditional Scottish materials if at all possible.
- Designs are to reflect a modern forward thinking Scotland.

Maintenance

- The information hub will provide easy maintenance and cleaning.
- The information hub will be durable, lasting for a period of five years.
- Replacement and repair should be considered in the proposal.

Cost

- Initially five units will be required. However, the potential to mass produce should be considered.
- The unit cost for each information hub should not exceed £3000.